Vladimir A. Tolstykh
Partial Differential Equations

I0054859

Also of Interest

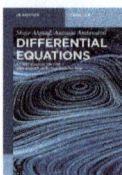

Differential Equations. A first course on ODE and a brief introduction to PDE
Shair Ahmad, Antonio Ambrosetti, 2019
ISBN 978-3-11-065003-7, e-ISBN (PDF) 978-3-11-065286-4,
e-ISBN (EPUB) 978-3-11-065008-2

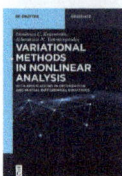

Variational Methods in Nonlinear Analysis. With Applications in Optimization and Partial Differential Equations
Dimitrios C. Kravvaritis, Athanasios N. Yannacopoulos, 2020
ISBN 978-3-11-064736-5, e-ISBN (PDF) 978-3-11-064738-9,
e-ISBN (EPUB) 978-3-11-064745-7

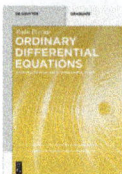

Ordinary Differential Equations. Example-driven, Including Maple Code
Radu Precup, 2018
ISBN 978-3-11-044742-2, e-ISBN (PDF) 978-3-11-044744-6,
e-ISBN (EPUB) 978-3-11-044750-7

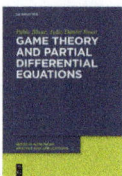

Game Theory and Partial Differential Equations
Pablo Blanc, Julio Daniel Rossi, 2019
ISBN 978-3-11-061925-6, e-ISBN (PDF) 978-3-11-062179-2,
e-ISBN (EPUB) 978-3-11-061932-4

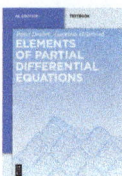

Elements of Partial Differential Equations
Pavel Drábek, Gabriela Holubová, 2014
ISBN 978-3-11-031665-0, e-ISBN (PDF) 978-3-11-031667-4,
e-ISBN (EPUB) 978-3-11-037404-9

Vladimir A. Tolstykh

Partial Differential Equations

——

An Unhurried Introduction

DE GRUYTER

Mathematics Subject Classification 2010
Primary: 35-01, 35A09, 35A24; Secondary: 35A30, 35-04

Author
Prof. Vladimir A. Tolstykh
Istanbul Arel University
Department of Mathematics
and Computer Science
Erguvan Sokak 26
34537 Istanbul
Turkey
vladimirtolstykh@arel.edu.tr

ISBN 978-3-11-067724-9
e-ISBN (PDF) 978-3-11-067725-6
e-ISBN (EPUB) 978-3-11-067733-1

Library of Congress Control Number: 2020934482

Bibliographic information published by the Deutsche Nationalbibliothek
The Deutsche Nationalbibliothek lists this publication in the Deutsche Nationalbibliografie;
detailed bibliographic data are available on the Internet at http://dnb.dnb.de.

© 2020 Walter de Gruyter GmbH, Berlin/Boston
Cover image: created by Vladimir Tolstykh with the use of PyPlot
Typesetting: VTeX UAB, Lithuania
Printing and binding: CPI books GmbH, Leck

www.degruyter.com

To the memory of my parents

Preface

This book originated as the lecture notes for a course in partial differential equations taught two years ago, quite by a chance and for the first time in my teaching career (I am an algebraist).

The topics covered in the book are the standard topics of a one-semester undergraduate course in partial differential equations (PDEs).

The first chapter introduces basic concepts and notation. The second chapter is devoted to the technique of change of variables in PDEs (judging by my own experience as a lecturer, the students often find the workings of the technique, which is first studied in multivariable calculus and almost never in sufficient detail, both mysterious and mystifying, whence the need to dispel the mist).

The third chapter deals with first-order linear PDEs with constant coefficients. The main topic of the fourth chapter is the method of characteristics for first-order semilinear PDEs.

First-order quasilinear equations are studied in Chapters 5–7: Chapter 5 introduces the subject, Chapter 6 is devoted to description of solution sets of first-order quasilinear PDEs, and Chapter 7 deals with the method of characteristics for this class of PDEs. Some parts of Chapters 5–7 may be considered as suitable for a graduate course in PDEs.

The main topic of the final Chapter 8 is second-order semilinear PDEs.

Throughout the book, we give numerous examples on the usage of Maple and the popular online resource Wolfram Alpha for solving problems related to the theory of PDEs. No prior knowledge of Maple, or prior experience of using Wolfram Alpha is required.

Added in proof. This book was in production during the murkiest days of the COVID-19 outbreak in Europe. It is with no small amount of admiration that I feel myself obliged to acknowledge the calm professionalism of the staff of my publisher De Gruyter and the typesetting company VTeX. My special thanks to Nadja Schedensack of De Gruyter and Ina Talandienė of VTeX.

Istanbul Arel University   *Vladimir A. Tolstykh*

https://doi.org/10.1515/9783110677256-201

Contents

Preface —— VII

1   Definition —— 1
1.1   Partial differential equations —— 1
1.2   Linear PDEs —— 4
1.3   Separable ODEs: a brief remainder —— 10
1.4   PDEs reducible to ODEs —— 16

2   Change of variables in PDEs —— 29
2.1   Change of variables in a PDE: "direct" approach —— 29
2.2   Inverse function theorem —— 39
2.3   Change of variables in a PDE: "indirect" approach —— 41

3   First-order linear equations —— 49

4   First-order semilinear equations —— 63
4.1   Peano–Pickard–Lindelöf theorem —— 63
4.2   Method of characteristics for semilinear equations —— 67

5   First-order quasilinear equations: vector fields —— 85
5.1   Peano–Pickard–Lindelöf theorem for higher dimensions —— 86
5.2   Vector fields and their characteristic curves —— 89
5.3   First integrals of vector fields —— 100

6   First-order quasilinear equations: solution sets —— 117
6.1   Solutions determined by first integrals —— 117
6.2   Main theorem on solution sets —— 124
6.3   Cauchy problem for a PDE —— 129
6.4   First integrals: up to the task once again —— 130

7   Method of characteristics for first-order quasilinear equations —— 157
7.1   Surfaces made up of characteristic curves —— 157
7.2   Main theorem —— 170

8   Second-order semilinear equations —— 205
8.1   Three main types —— 205
8.2   Hyperbolic equations —— 214
8.3   Parabolic equations —— 223
8.4   Elliptic equations —— 231

A **Appendix** —— **249**
A.1 Inverse function theorem —— **249**
A.2 Functional dependence —— **256**

Bibliography —— 263

Index —— 265

1 Definition

1.1 Partial differential equations

Recall that an *ordinary differential equation* (ODE) for an unknown function $u = u(x)$ in one variable can be (re)written in the form

$$F(x, u(x), u'(x), u''(x), \ldots) = 0,$$

or, for short, as

$$F(x, u, u', u'', \ldots) = 0,$$

where $F(x, y_0, y_1, y_2, \ldots)$ is a function in independent variables $x, y_0, y_1, y_2, \ldots$

Let $u = u(x, y, \ldots)$ be a function/variable which depends on *more than one* variable. An equation

$$F(x, y, \ldots, u, u_1^{(1)}, u_2^{(1)}, \ldots, u_1^{(2)}, u_2^{(2)}, \ldots, u_1^{(n)}, u_2^{(n)}, \ldots) = 0,$$

where, for the sake of notational simplicity, the symbols $u_i^{(k)}$ denote the partial derivatives of u order k, is called a *partial differential equation* (PDE).

Following a very convenient notation from calculus, we shall always denote the partial derivatives

$$\frac{\partial u}{\partial x}, \frac{\partial u}{\partial y}, \frac{\partial u}{\partial z}, \ldots$$

of a function u by $x, y, z, \ldots$ by

$$u_x, u_y, u_z, \ldots$$

respectively, the second-order partial derivatives of u,

$$\frac{\partial^2 u}{\partial x^2}, \frac{\partial^2 u}{\partial x \partial y}, \frac{\partial^2 u}{\partial y \partial x}, \frac{\partial^2 u}{\partial y^2}$$

are then denoted by

$$u_{xx}, u_{xy}, u_{yx}, u_{yy},$$

respectively, and so on.

For instance, the general PDE for an unknown function $u = u(x, y)$, which contains partial derivatives of u of at most first order can be written as

$$F(x, y, u(x, y), u_x(x, y), u_y(x, y)) = 0, \tag{1.1.1}$$

https://doi.org/10.1515/9783110677256-001

or, for short, as

$$F(x, y, u, u_x, u_y) = 0.$$

As in most of the introductory books on the subject, we shall mainly restrict ourselves to the case of PDEs in *two independent variables* (much like when the student reads multivariable calculus for the first time).

The *order* of a partial differential equation is the highest order of the partial derivative that appears in the equation.

A *solution* of a PDE of order n on an open set $\Omega \subseteq \mathbf{R}^2$ is an n times continuously differentiable function $u = u(x, y)$ defined on Ω, symbolically, $u \in C^n(\Omega)$, which satisfies this PDE for all points $\begin{pmatrix} x \\ y \end{pmatrix} \in \Omega$.

For instance, given a PDE of the form

$$a(x, y)u_x + b(x, y)u_y + c(x, y)u = d(x, y),$$

we can consider as Ω *any* open subset of the common domain of the functions $a(x, y)$, $b(x, y)$, $c(x, y)$, and $d(x, y)$, or, in other words, Ω can be any open set in $\mathbf{R}^2$ satisfying

$$\Omega \subseteq \operatorname{dom}(a) \cap \operatorname{dom}(b) \cap \operatorname{dom}(c) \cap \operatorname{dom}(d).$$

Quite often, Ω is required be not only open but also connected, that is, be a *domain* in $\mathbf{R}^2$ (for instance, an open disk in $\mathbf{R}^2$). The reader may recollect that usually solutions of ODEs are first analyzed on open intervals in $\mathbf{R}$, open and connected sets in $\mathbf{R}$, and, following this, on other open sets in $\mathbf{R}$.

As is customary, we take as our first examples a number of PDEs one meets when studying various physical disciplines:

$u_x + u_y = 0$ (transport equation, a first-order PDE),

$u_t + uu_x = 0$ (inviscid Burgers' equation, a first-order PDE),

$u_{xx} + u_{yy} = 0$ (Laplace's equation, a second-order PDE),

$u_{tt} - u_{xx} = 0$ (wave equation, a second-order PDE),

$u_t - u_{xx} = 0$ (heat equation, a second-order PDE),

$u_t + uu_x + u_{xxx} = 0$ (Korteweg–de Vries, or KdV, equation, a third-order PDE).

Let us demonstrate that the function

$$u(t, x) = \cos(t + x)$$

is a solution of the wave equation $u_{tt} - u_{xx} = 0$ on $\Omega = \mathbf{R}^2$: to do so, we simply substitute this function into the equation:

$$\left[\cos(t + x)\right]_{tt} - \left[\cos(t + x)\right]_{xx} = -\cos(t + x) - (-\cos(t + x)) = 0$$

for all $t, x \in \mathbf{R}$.

Using mathematical software can, as the reader undoubtedly knows, ease up the process of studying most of mathematical disciplines, with differential equations being no exception. Time to time, we shall show how to deal with problems involving PDEs using the well-known online resource Wolfram Alpha ($W|\alpha$) and the popular program package Maple.

For instance, the following query:

```
D[ u(t,x), t,t ] - D[ u(t,x), x,x ]=0
```

entered at the main webpage of Wolfram Alpha (https://wolframalpha.com) provides the general solution of the wave equation given as

$$u(t, x) = c_1(t - x) + c_2(t + x),$$

where c_1, c_2 are arbitrary twice continuously differentiable functions defined everywhere on $\mathbf{R}$ (note that the solution $u(t, x) = \cos(t + x)$ we have considered above is of this form).

The corresponding query/command in Maple is as follows:

```
pdsolve( diff( u(t,x), t,t ) - diff( u(t,x), x,x ) );
```

and it produces the (same) general solution of the wave equation given as

$$u(t, x) = \_F1(x + t) + \_F2(x - t),$$

where, again, _F1, _F2 are twice continuously differentiable functions that are defined on $\mathbf{R}$.

The reader is encouraged to figure out how to form the queries/commands to get the general solutions (if available) for all PDEs given above with the use of $W|\alpha$, or Maple. Generally speaking, the (very encouraging!) fact that a particular PDE can be handled by mathematical software usually means that there is a well-developed theory behind a certain class of PDEs into which the equation can be included. More often than not, however, no mathematical software can help in obtaining general solutions of PDEs, but then there exist methods for obtaining particular solutions, and, last but not least, there are powerful methods for obtaining numerical solutions, etc.

1.2 Linear PDEs

In this section, as is also customary, we shall discuss the definition and outline some basic properties of the linear PDEs – in order to illustrate the definition of a PDE. However, to smooth the discussion, we shall start with repeating some main points of the theory of linear *ODEs*.

As the reader may remember, the general representation of an ODE in the form

$$F(x, u, u', u'', \ldots, u^{(n)}) = 0 \tag{1.2.1}$$

is fairly rarely used. Well, except for the cases when one introduces a new class of ODEs. Let O be an open set in $\mathbf{R}$ and let

$$a_0(x), a_1(x), \ldots, a_n(x), g(x)$$

be functions which are continuous on O. Then the ODE

$$a_0(x)u + a_1(x)u' + \cdots + a_n(x)u^{(n)} = g(x) \tag{1.2.2}$$

is called *linear*. The term is due to the fact that the function

$$F(x, y_0, y_1, \ldots, y_n) = a_0(x)y_0 + a_1(x)y_1 + \cdots + a_n(x)y_n - g(x),$$

which can be used to represent the equation (1.2.2) in the general form (1.2.1), is linear with respect to the variables $y_0, y_1, \ldots, y_n$. Linear algebra, one of the most helpfully formal mathematical disciplines, can get the things straight: the mentioned property of the variables $y_0, y_1, \ldots, y_n$ means that equivalently, in the sense of linear algebra, the function

$$\Phi(y_0, y_1, \ldots, y_n) = a_0(x)y_0 + a_1(x)y_1 + \cdots + a_n(x)y_n$$

is a linear map from the vector space

$$C(O)^{n+1} = \underbrace{C(O) \times C(O) \times \cdots \times C(O)}_{(n+1)\ \text{times}}$$

into the vector space $C(O)$, that is,

$$\Phi(\vec{v} + \vec{w}) = \Phi(\vec{v}) + \Phi(\vec{w})$$

and

$$\Phi(\alpha\vec{v}) = \alpha\Phi(\vec{v}),$$

where

$$\vec{v} = (v_0(x), v_1(x), \ldots, v_n(x)) \quad \text{and} \quad \vec{w} = (w_0(x), w_1(x), \ldots, w_n(x))$$

are arbitrary tuples of continuous functions in $C(O)^{n+1}$ and $\alpha \in \mathbf{R}$ is an arbitrary scalar (proof?). It follows that the function

$$\mathcal{L}(u) = a_0(x)u + a_1(x)u' + \cdots + a_n(x)u^{(n)}$$

is a linear map from the vector space $C^n(O)$ into the vector space $C(O)$, and hence

$$\mathcal{L}(u + v) = \mathcal{L}(u) + \mathcal{L}(v),$$
$$\mathcal{L}(\alpha u) = \alpha \mathcal{L}(u)$$

for all $u, v \in C^n(O)$. The linear equation (1.2.2) can therefore be written as

$$\mathcal{L}(u) = g(x),$$

and is called *homogeneous* if $g(x)$ is the zero function and *inhomogeneous*, otherwise.

Next, we shall quote an important structural result for the solutions sets of linear ODEs (whose analog is also true for linear PDEs, as we shall see below).

Proposition 1.2.1. (i) *The set of solutions of a homogeneous linear ODE is a subspace of the corresponding space of functions.*

(ii) *Given an inhomogeneous linear ODE*

$$a_n(x)u^{(n)} + a_{n-1}(x)u^{(n-1)} + \cdots + a_0(x)u = g(x), \tag{1.2.3}$$

its general solution (if exists) can be written in the form

$$u = u_{\mathrm{prt}} + u_{\mathrm{hom}},$$

where u_{prt} is a particular solution of the equation (1.2.3) and u_{hom} is the general solution of the corresponding homogeneous equation

$$a_n(x)u^{(n)} + a_{n-1}(x)u^{(n-1)} + \cdots + a_0(x)u = 0.$$

Now back to PDEs. A *linear* PDE is one which can be written in the form

$$a_0(x,y)u + a_1(x,y)u_x + a_2(x,y)u_y + \cdots = g(x,y), \tag{1.2.4}$$

where each term of the sum in the left-hand side is a product of a function $a_i(x,y)$ and a partial derivative of the unknown function $u = u(x,y)$; the coefficient functions $a_i(x,y)$ and the function $g(x,y)$ are all assumed to be continuous on some open set $\Omega \subseteq \mathbf{R}^2$. As in the case with ODEs, the term is due to the fact the equation can be written in the general form

$$F(x,y,u,u_1^{(1)},u_2^{(1)},u_1^{(2)},u_2^{(2)},\ldots,u_1^{(n)},u_2^{(n)},\ldots) = 0,$$

(recall that the symbols $u_i^{(k)}$ are used to denote the partial derivatives of u of order k) by means of a function

$$F(x, y, z_0, z_1^{(1)}, z_2^{(1)}, z_1^{(2)}, z_2^{(2)}, \ldots, z_1^{(n)}, z_2^{(n)}, \ldots),$$

which is linear with regard to the variables

$$z_0, z_1^{(1)}, z_2^{(1)}, \ldots, z_1^{(2)}, z_2^{(2)}, \ldots, z_1^{(n)}, z_2^{(n)}, \ldots \tag{1.2.5}$$

More precisely, we should have that

$$F(x, y, \vec{z}) = \Phi(\vec{z}) - g(x, y),$$

where $\vec{z}$ is the tuple of variables listed in (1.2.5) and the function Φ is of the form

$$\Phi(\vec{z}) = \sum_{t \in \vec{z}} a_t(x, y)t,$$

where $a_t(x, y) \in C(\Omega)$ for each variable t in $\vec{z}$, and hence it determines a linear map from the vector space $C(\Omega)^{|\vec{z}|}$ into the vector space $C(\Omega)$.

For instance, a PDE

$$a(x, y)u_x + b(x, y)u_y + c(x, y)u = d(x, y),$$

where the coefficients functions

$$a(x, y), b(x, y), c(x, y)$$

and the function $d(x, y)$ are all continuous on an open set Ω in $\mathbf{R}^2$, is a linear PDE of order at most one. Observe that the coefficients functions depend solely on the independent variables x, y and, say no product of partial derivatives of u enters the equation. The following function:

$$F(x, y, z_0, z_1^{(1)}, z_2^{(1)}) = c(x, y)z_0 + a(x, y)z_1^{(1)} + b(x, y)z_2^{(1)} - d(x, y),$$

which is linear with regard to the variables

$$z_0, z_1^{(1)}, \text{ and } z_2^{(1)},$$

can be used to represent this equation in the form

$$F(x, y, u, u_x, u_y) = 0.$$

Suppose that the linear PDE (1.2.4) is of order n. Then it follows from the definition that the left-hand side of (1.2.4),

$$\mathcal{L}(u) = a_0(x, y)u + a_1(x, y)u_x + a_2(x, y)u_y + \cdots$$

determines a linear map from the vector space $C^n(\Omega)$ into the vector space $C(\Omega)$ of continuous functions on Ω: the linearity properties

$$\mathcal{L}(u + v) = \mathcal{L}(u) + \mathcal{L}(v), \qquad (1.2.6)$$
$$\mathcal{L}(\alpha u) = \alpha \mathcal{L}(u)$$

are true all $u, v \in C^n(\Omega)$ and for all $\alpha \in \mathbf{R}$. Furthermore, the PDE (1.2.4) can be written in the form

$$\mathcal{L}(u) = g(x, y),$$

and in the case when $g(x, y)$ is the zero function, the linear PDE (1.2.4) is called *homogeneous*, and *inhomogeneous*, otherwise.

For instance, the linear equation

$$u + u_{xx} = 0$$

is a homogeneous, whereas the linear equation

$$u + u_{xx} = x^2 - y^2$$

is an inhomogeneous linear equation.

Further examples with the equations above:

$$u_x + u_y = 0 \qquad \text{(transport equation, linear)},$$
$$u_t + u u_x = 0 \qquad \text{(Burgers' equation, nonlinear)},$$
$$u_{xx} + u_{yy} = 0 \qquad \text{(Laplace's equation, linear)}$$
$$u_{tt} - u_{xx} = 0 \qquad \text{(wave equation, linear)},$$
$$u_t - u_{xx} = 0 \qquad \text{(heat equation, linear)},$$
$$u_t + u u_x + u_{xxx} = 0 \qquad \text{(KdV equation, nonlinear)}.$$

To make sure that a certain PDE is *not* linear, one can first rewrite in the form

$$\mathcal{L}(u) = g(x, y)$$

and then demonstrate that the corresponding map $\mathcal{L}$ fails to satisfy one of the linearity properties in (1.2.6). For example, let us justify the above claim that Burgers' equation

$$u_t + u u_x = 0$$

on $\Omega = \mathbf{R}^2$ is not linear. We have to show that the map

$$\mathcal{L}(u) = u_t + u u_x$$

from $C^1(\mathbf{R}^2)$ into $C(\mathbf{R})$, determined by the equation, is not linear. To show this, let us consider the function

$$u = u(t,x) = x$$

on $\mathbf{R}^2$. Then, on one hand,

$$\mathcal{L}(2u) = \mathcal{L}(2x) = [2x]_t + (2x)[2x]_x = 4x,$$

while, on the other hand,

$$2\mathcal{L}(u) = 2\mathcal{L}(x) = 2([x]_t + x[x]_x) = 2x,$$

and then

$$\mathcal{L}(2u) \neq 2\mathcal{L}(u).$$

So the map $\mathcal{L}$ does not preserve the scalar multiplication (the second property in (1.2.6), which, as we have demonstrated above, fails for the vector $u = x$ and the scalar $\alpha = 2$), and hence $\mathcal{L}$ is not linear, as claimed. Note that our argument also implies that $\mathcal{L}$ does not preserve the addition as well, since

$$\mathcal{L}(2u) = \underbrace{\mathcal{L}(u + u) \neq \mathcal{L}(u) + \mathcal{L}(u)} = 2\mathcal{L}(u).$$

Proposition 1.2.2. *Consider a homogeneous linear PDE*

$$a_0(x,y)u + a_1(x,y)u_x + a_2(x,y)u_y + \cdots = 0 \tag{1.2.7}$$

of order n on an open set $\Omega \subseteq \mathbf{R}^2$, and let

$$\mathcal{L}(u) = a_0(x,y)u + a_1(x,y)u_x + a_2(x,y)u_y + \cdots$$

be the corresponding linear map from $C^n(\Omega)$ into $C(\Omega)$. Then:
   (i) *the set of solutions of the homogeneous PDE (1.2.7) is a subspace of the corresponding space of functions $C^n(\Omega)$;*
   (ii) *if*

$$\mathcal{L}(u) = g(x,y), \tag{1.2.8}$$

where g is a continuous function on Ω, then the general solution of the equation (1.2.8) (if exists) can be written as

$$u = u_{\mathrm{prt}} + u_{\mathrm{hom}},$$

where u_{prt} is a particular solution of the equation (1.2.8) and u_{hom} is the general solution of the correspond homogeneous linear equation (1.2.7).

Proof. (i) Let $u, v \in C^n(\Omega)$ be solutions of the homogeneous equation:

$$\mathcal{L}(u) = \mathcal{L}(v) = 0.$$

Then, due to linearity of $\mathcal{L}$,

$$\mathcal{L}(u + v) = \mathcal{L}(u) + \mathcal{L}(v) = 0 + 0 = 0 \Rightarrow \mathcal{L}(u + v) = 0$$

and hence $u + v$ is also a solution of the homogeneous equation.

Further, for every scalar $\alpha \in \mathbf{R}$,

$$\mathcal{L}(\alpha u) = \alpha \mathcal{L}(u) = \alpha 0 = 0$$

and αu is again a solution of our PDE.

(ii) Now consider the inhomogeneous linear PDE

$$\mathcal{L}(u) = g(x, y)$$

whose solution set is nonempty, and hence contains a function $u_{\text{prt}} \in C^n(\Omega)$.

If a function $u \in C^n(\Omega)$ can be written as

$$u = u_{\text{prt}} + u_{\text{hom}},$$

where u_{hom} is a solution of the corresponding homogeneous equation $\mathcal{L} = 0$, then due to (once again) linearity of $\mathcal{L}$, u is a solution of the equation (1.2.8):

$$\begin{aligned} \mathcal{L}(u) = \mathcal{L}(u_{\text{prt}} + u_{\text{hom}}) &= \mathcal{L}(u_{\text{prt}}) + \mathcal{L}(u_{\text{hom}}) \\ &= g(x, y) + 0 \\ &= g(x, y). \end{aligned}$$

Conversely, let $u \in C^n(\Omega)$ be an arbitrary solution of the equation (1.2.8):

$$\mathcal{L}(u) = g(x, y).$$

It then follows that

$$\begin{aligned} \mathcal{L}(u - u_{\text{prt}}) &= \mathcal{L}(u) - \mathcal{L}(u_{\text{prt}}) \\ &= g(x, y) - g(x, y) \\ &= 0, \end{aligned}$$

and hence $u - u_{\text{prt}}$ is a solution of the corresponding homogeneous equation. So

$$u = u_{\text{prt}} + (u - u_{\text{prt}}) = u_{\text{prt}} + u_{\text{hom}},$$

which completes the proof. □

As the reader may remember from the course in ODEs, the solution set of a homogeneous linear ODE of order n on an open interval I in $\mathbf{R}$ is (under certain natural assumptions, "normally") a subspace of dimension n in the corresponding space of functions, $C^n(I)$. As far as homogenous linear PDEs are concerned, their solutions sets are "normally" *infinite-dimensional* subspaces in the corresponding spaces of functions. For instance, as we have seen above, the solution set S of the wave equation $u_{tt} - u_{xx} = 0$ on $\mathbf{R}^2$ is given by

$$\{f(x + t) + g(x - t) : f, g \in C^2(\mathbf{R})\}.$$

Now since, say, the functions

$$1, (x + t), (x + t)^2, \ldots, (x + t)^n, \ldots,$$

where $n \in \mathbf{N}$, in S are linearly independent, S is an infinite-dimensional subspace of the vector space $C^2(\mathbf{R}^2)$.

1.3 Separable ODEs: a brief remainder

As we shall see below, first-order PDEs are often *reducible* to ODEs (which means that once a suitable related ODE is solved, the original PDE can be also solved). Due to that, we shall repeat some basic facts on ODEs that will be required below.

An important, and a mostly-easy-to-deal-with, class of ODEs is the class of so-called

> *separable* ODEs

(we recall the definition shortly).

Recall that given a real-valued function $\varphi : I \to \mathbf{R}$ on an open interval in $\mathbf{R}$, the set

$$\text{Graph}(\varphi) = \left\{ \begin{pmatrix} x \\ \varphi(x) \end{pmatrix} : x \in I \right\}$$

is called the *graph* of φ. The key statement on a class of ODEs which *contains* the class of separable ODEs is as follows.

Proposition 1.3.1 (Exact differential equations). *Let Ω be a domain in $\mathbf{R}^2$. Suppose that given an ODE*

$$y' = \frac{b(x, y)}{a(x, y)}, \tag{1.3.1}$$

where $a(x, y), b(x, y) \in C(\Omega)$ and $a(x, y) \neq 0$ everywhere on Ω, there is a continuously differentiable function $F : \Omega \to \mathbf{R}$ such that

$$F_x(x, y) = b(x, y) \quad and \quad F_y(x, y) = -a(x, y)$$

everywhere on Ω. Consequently, the equation (1.3.1) *can be rewritten as*

$$- F_y(x,y)dy = F_x(x,y)dx \iff d(F(x,y)) = 0. \tag{1.3.2}$$

Let then $\varphi : I \to \mathbf{R}$ *be a continuously differentiable function on an open interval I in* **R** *whose graph lies entirely in* Ω. *Then* φ *is a solution of* (1.3.2) *if and only if*

$$F(x, \varphi(x)) = C = \text{const}$$

for all $x \in I$.

Proof. Observe that if the graph of a smooth function $\varphi : I \to \mathbf{R}$ is contained in Ω, then by the conditions,

$$\begin{aligned}[F(x, \varphi(x))]_x &= F_x(x, \varphi(x)) + F_y(x, \varphi(x))\varphi'(x) \\ &= b(x, \varphi(x)) - a(x, \varphi(x))\varphi'(x)\end{aligned} \tag{1.3.3}$$

for all $x \in I$.

($\Rightarrow$). Suppose that $\varphi(x)$ is a solution of the ODE (1.3.1), that is,

$$\varphi'(x) = \frac{b(x, \varphi(x))}{a(x, \varphi(x))}$$

for all $x \in I$. Then

$$b(x, \varphi(x)) - a(x, \varphi(x))\varphi'(x) = 0$$

for all $x \in I$, whence, by (1.3.3),

$$[F(x, \varphi(x))]_x = 0$$

for all $x \in I$. It follows that the function $F(x, \varphi(x))$ is constant on I:

$$F(x, \varphi(x)) = C = \text{const}$$

for all $x \in I$.

($\Leftarrow$). Conversely, suppose that

$$F(x, \varphi(x)) = C = \text{const}$$

for all $x \in I$. We differentiate both parts by x, thereby getting, again by (1.3.3), that

$$0 = [F(x, \varphi(x))]_x = b(x, \varphi(x)) - a(x, \varphi(x))\varphi'(x),$$

for all $x \in I$, or

$$\varphi'(x) = \frac{b(x, \varphi(x))}{a(x, \varphi(x))}$$

for all $x \in I$. Thus, φ is a solution of the ODE in question, as required. ☐

The reader can recollect that Proposition 1.3.1 can be conveniently restated in the following form: a function $F \in C^1(\Omega)$ which satisfies (1.3.2) is a first integral of the ODE (1.3.1) on Ω (if the reader is unfamiliar with the term, it can be accepted "as-is," but since first integrals of explicit ODEs $y' = f(x,y)$ will play an important role in Chapter 4, we certainly reproduce the corresponding definition in the beginning of the chapter).

Two continuous functions $b(x) \in C(I_0)$ and $a(y) \in C(J_0)$, where I_0, J_0 are open intervals in **R**, and where

$$a(y) \neq 0$$

for all $y \in J_0$, determine the ODE

$$y' = \frac{b(x)}{a(y)}, \tag{1.3.4}$$

which is called *separable*. Understandably, any solution of this ODE must be a continuously differentiable function $y = \varphi(x) \in C^1(I)$, where I is an open subinterval of I_0, whose range is contained in J_0:

$$I = \text{dom}(\varphi) \subseteq I_0 \quad \text{and} \quad \text{rng}(\varphi) \subseteq J_0.$$

Further, since the function $a(y)$ is nonzero everywhere on J_0,

$$y' = \frac{b(x)}{a(y)} \iff a(y)y' = b(x)$$

$$\iff a(y)dy = b(x)dx.$$

In the last equation above,

$$a(y)dy = b(x)dx, \tag{1.3.5}$$

the variables x and y are *separated* from each other, whence the term.

We wish to apply Proposition 1.3.1 in order to obtain a first integral of the equation (1.3.5). To this goal, consider the function

$$F(x,y) = \int b(x)\,dx - \int a(y)\,dy.$$

Clearly,

$$F_x(x,y) = \left[\int b(x)\,dx - \int a(y)\,dy \right]_x$$

$$= \left[\int b(x)\,dx \right]_x$$

$$= b(x)$$

and

$$F_y(x,y) = \left[\int b(x)\,dx - \int a(y)\,dy\right]_y$$
$$= \left[-\int a(y)\,dy\right]_y$$
$$= -a(y).$$

Hence the function F satisfies the conditions of Proposition 1.3.1 for $\Omega = I_0 \times J_0$, and so it is indeed a first integral of the ODE (1.3.4) on Ω. Strictly formally, we then get a *description* of the general solution of the ODE (1.3.4) in the form

$$F(x,y) = C,$$

where C is a constant, meaning that a smooth function $\varphi : I \rightarrow \mathbf{R}$, where $I \subseteq I_0$ is an open interval, whose range is contained in J_0, is a solution of the ODE (1.3.4) if and only if φ satisfies the functional equation

$$F(x,y(x)) = C$$

for a suitable constant $C \in \mathbf{R}$ everywhere on I.

In practice, if one suspects that a certain ODE is separable, one first tries to rewrite it in the form (1.3.4) and then in the form (1.3.5); following this, one integrates both parts in order to obtain a first integral of the ODE.

Example 1.3.1. Let $\lambda \in \mathbf{R}$. Then the general solution of the ODE

$$y' = \lambda y$$

on $\mathbf{R}$ is

$$y = Ce^{\lambda x},$$

where C is an arbitrary constant.

Solution. The variables are already nearly separated:

$$y' = \lambda y \iff \frac{dy}{dx} = \lambda y$$

They *will* be separated if we move y to the left-hand side, that is, divide both parts by y.

This means that we have to *assume* that, keeping in mind the definition of a separable ODE, while the variable x can vary in any open interval I_0 in $\mathbf{R}$, the variable y must vary in an open interval J_0 which does *not* contain 0.

Let us proceed under this assumption. Then

$$y' = \lambda y \Rightarrow \frac{dy}{y} = \lambda dx.$$

Integrating both parts, we get that

$$\int \frac{dy}{y} = \int \lambda dx \Rightarrow \ln(|y|) = \lambda x + C_0$$

where C_0 is an arbitrary constant. Applying the exponential function to both sides of the last equation, we get that

$$\exp(\ln(|y|)) = |y| = \exp(\lambda x + C_0) \Rightarrow |y| = e^{C_0} e^{\lambda x}.$$

It follows that

$$y = \pm e^{C_0} e^{\lambda x}.$$

As y is differentiable, y is *continuous*, which implies that

$$y = e^{C_0} e^{\lambda x}, \quad \text{or} \quad y = -e^{C_0} e^{\lambda x},$$

or, in other words,

$$y = Ce^{\lambda x},$$

where C is an arbitrary constant.

So if y is continuously differentiable, satisfies the ODE

$$y' = \lambda y,$$

and *nonzero* on a certain interval I_0 in **R**, then

$$y = Ce^{\lambda x}$$

on this interval. Now such a function y can be *extended* to the function

$$y = Ce^{\lambda x}$$

on the whole of **R** (**R** being the largest possible domain).

Next, the zero function on **R** is also of the form $Ce^{\lambda x}$ and satisfies our equation. Then we can conclude

$$y' = \lambda y \Rightarrow y = Ce^{\lambda x}$$

everywhere on **R**, where C is an arbitrary constant.

Conversely, any function of the above form is a solution of our ODE:

$$(Ce^{\lambda x})' = C\lambda e^{\lambda x} = \lambda(Ce^{\lambda x}).$$

Thus

$$y' = \lambda y \iff y = Ce^{\lambda x},$$

as claimed. $\qquad\square$

Example 1.3.2. Solve the ODE

$$xy' = y$$

on **R**.

Solution. We have that

$$x\frac{dy}{dx} = y.$$

In order to separate the variables, we have to divide by x and by y. What does it mean?

Well, it means that we have to assume that *each* of the variables x, y varies in an open interval which does not contain 0.

Let us proceed, remembering this assumption:

$$\frac{dy}{y} = \frac{dx}{x} \Rightarrow \int \frac{dy}{y} = \int \frac{dx}{x}$$

whence

$$\ln(|y|) = \ln(|x|) + C_0$$

Applying the exponential function to both parts, we get that

$$|y| = |x| \exp(C_0) \Rightarrow \pm y = \pm Cx \Rightarrow y = Cx.$$

So what do we have? *Under our assumption,*

$$xy' = y \Rightarrow y = Cx,$$

where C is an arbitrary nonzero constant. As in the previous example, any such solution can be extended to the whole of **R**, the largest possible domain, and the zero function on **R** is also of the form

$$y = Cx$$

(as in the previous example, in the situation at hand everything is fine, but that is where one must be careful making sure that there is no clash with the assumptions).

The converse is certainly true:

$$y = Cx \Rightarrow xy' = x(Cx)' = Cx \Rightarrow xy' = y.$$

Therefore,

$$xy' = y \iff y = Cx,$$

where C is an arbitrary constant, is the general solution of our ODE on **R**. □

Let us compare our solution with the solutions provided by $W|\alpha$ and Maple. The corresponding query to $W|\alpha$ can be as simple as

```
xy'=y
```

which produces the solution given as

$$y = c_1 x$$

(and c_1 is to be treated as an arbitrary constant, of course). The solution with Maple is obtained with the use of the command

```
dsolve( x*diff(y(x), x) = y(x) );
```

which produces

$$y(x) = \_C1 x.$$

Note that sometimes it may be safer to use the command `dsolve` in the form

```
dsolve( ODE , y(x) );
```

that is, providing a reference to the unknown function of the ODE.

1.4 PDEs reducible to ODEs

In this section, we shall work out in detail solutions of several simple PDEs that are reducible to ODEs.

Example 1.4.1. Find the general solution of the PDE

$$u_x = 0$$

on $\Omega = \mathbf{R}^2$.

A couple of words before we turn on to the solution of the example. Some people have trouble understanding that the subscript x in the notation u_x is merely a reference to the *first* variable/argument of the function $u = u(x, y)$. To help with this, some textbooks use the notation like

$$u_1, u_2, u_{1,1}, u_{1,2}, u_{2,1}, u_{2,2} \dots$$

Thus for functions

$$F = F(x, y) \quad \text{and} \quad G = G(s, t),$$

the symbols

$$F_1 \quad \text{and} \quad G_1$$

will refer to the partial derivatives with regard to the first variable, F_x and G_s, respectively, and the symbols

$$F_{2,1} \quad \text{and} \quad G_{2,1}$$

to the second-order partial derivatives with respect to the second and to the first variable, that is,

$$F_{2,1} = F_{yx} \quad \text{and} \quad G_{2,1} = G_{ts}.$$

In particular, both the PDE

$$u_x = 0$$

for a C^1-function $u = u(x, y)$ on $\mathbf{R}^2$, and say the PDE

$$w_s = 0$$

for a C^1-function $w = w(s, t)$ on $\mathbf{R}^2$, can be uniformly represented in the form

$$u_1 = 0 \quad \text{and} \quad w_1 = 0,$$

and solutions of both these equations are exactly C^1-functions on $\mathbf{R}^2$ whose partial derivative with respect to the first argument is identically zero on $\mathbf{R}^2$. That said, expressions like

$$u_x(x^2, y - x)$$

are surely require some (small) time to get accustomed to, but the alternative expressions like

$$u_1(x^2, y - x), \quad \text{or} \quad D_1 u(x^2, y - x)$$

are lacking in charm and tradition.

Solution of Example 1.4.1. Assume that a continuously differentiable function $u = u(x, y)$ on $\mathbf{R}^2$ is a solution of the PDE

$$u_x = 0$$

on $\mathbf{R}^2$. As was discussed above, the partial derivative of u with respect to the first argument is then a function which is identically zero on $\mathbf{R}^2$:

$$u_x(\mathbf{x}, \mathbf{y}) = 0$$

for *all* $\begin{pmatrix} x \\ y \end{pmatrix} \in \mathbf{R}^2$. In particular, for every number $y_0 \in \mathbf{R}$,

$$u_x(x, y_0) = 0,$$

for all points of the form $\begin{pmatrix} x \\ y_0 \end{pmatrix} \in \mathbf{R}^2$. Consequently,

$$[u(x, y_0)]_x = [u(x, y_0)]' = 0$$

for all $x \in \mathbf{R}$, which implies that the function

$$x \mapsto u(x, y_0)$$

in the variable x is constant:

$$u(x, y_0) = \text{const}$$

for all $x \in \mathbf{R}$. It is quite clear that as y_0 *changes*, the corresponding constant also *changes*, and we can express this by "saying" that

$$u(x, y_0) = f(y_0),$$

for all $x, y_0 \in \mathbf{R}$. But then

$$u(x, y) = f(y),$$

where f is a continuously differentiable function.

Therefore,

$$u_x = 0 \Rightarrow u(x, y) = f(y).$$

The reader is urged to remember that when aiming at getting the *general* solution of a given PDE, our ultimate goal is to obtain a *statement* of the form

$$F(x, u, u_x, u_y, \ldots) = 0$$
$$\Updownarrow$$
$$u(x, y) = f(x, y), \quad (f \in \mathcal{F}),$$

where $\mathcal{F}$ is a certain family of functions.

Quite frequently, we first do the necessity part, that is, *prove* (demonstrate) that

$$F(x, u, u_x, u_y, \ldots) = 0$$
$$\Downarrow$$
$$u(x, y) = f(x, y), \quad (f \in \mathcal{F}),$$

and then we *verify* that *any* function $f \in \mathcal{F}$ satisfies the original PDE, thereby getting that

$$F(x, u, u_x, u_y, \ldots) = 0$$
$$\Uparrow$$
$$u(x, y) = f(x, y), \quad (f \in \mathcal{F}),$$

and, finally, that

$$F(x, u, u_x, u_y, \ldots) = 0$$
$$\Updownarrow$$
$$u(x, y) = f(x, y), \quad (f \in \mathcal{F}).$$

Now, getting back to our problem, we ask ourselves whether the converse,

$$u_x = 0 \Leftarrow u(x, y) = f(y),$$

is true?

It certainly is:

$$u(x, y) = f(y) \Rightarrow u_x = [f(y)]_x = 0 \Rightarrow u_x = 0.$$

Therefore, the general solution of our PDE on $\mathbf{R}^2$ is

$$u_x = 0 \iff u(x, y) = f(y),$$

where f is a continuously differentiable function on $\mathbf{R}$. □

To solve the last equation with Maple: enter the command

```
pdsolve( diff( u(x,y), x) = 0 );
```

which produces

$$u(x, y) = \_F1(y);$$

the corresponding query to $\mathcal{W}\!\alpha$ is

```
D[ u(x,y), x ]=0
```

which produces

$$u(x, y) = c_1(y).$$

Example 1.4.2. Find the general solution of the PDE

$$u_{xx} + u = 0$$

on $\mathbf{R}^2$.

Solution. Similar to the previous example, the PDE contains only partial derivatives *with respect to x* (there are *no* partial derivatives with respect y):

$$[u(x,y)]_{xx} + u(x,y) = 0, \tag{1.4.1}$$

and (1.4.1) is true for *all* $\begin{pmatrix} x \\ y \end{pmatrix} \in \mathbf{R}^2$.

Then we again substitute some fixed value $y_0 \in \mathbf{R}$ for y, thereby getting that

$$[u(x,y_0)]_{xx} + u(x,y_0) = 0 \iff [h(x)]_{xx} + h(x) = 0,$$

and so the function $h(x) = u(x,y_0)$ is a solution of the ODE

$$z''(x) + z(x) = 0.$$

As we (have to) know, the ODE

$$z_{xx} + z = 0 \iff z'' + z = 0$$

has the general solution

$$z(x) = C_1 \cos x + C_2 \sin x,$$

where C_1, C_2 are arbitrary constants (if this result is forgotten, the reader is advised to review the linear ODEs with constant coefficients; at the very least, query W|a: a sufficiently effective math software can be viewed as a giant "handbook" which may be consulted if needed).

Thus

$$h(x) = u(x,y_0) = f(y_0) \cos x + g(y_0) \sin x,$$

where f and g are differentiable functions,

since each $y_0 \in \mathbf{R}$ "gets" its own constants C_1, C_2; as y_0 changes, the constants change as well.

Switching from y_0 back to y, we get that

$$u(x,y) = f(y) \cos x + g(y) \sin x.$$

To summarize,

$$u_{xx} + u = 0 \Rightarrow u(x,y) = f(y) \cos x + g(y) \sin x,$$

where $f, g : \mathbf{R} \to \mathbf{R}$ are twice continuously differentiable functions (why "twice"? since any solution $u(x,y)$ is assumed to be a twice continuously differentiable function on $\mathbf{R}^2$). Again, we have to ask ourselves:

is the converse true?

It is: if

$$u(x,y) = f(y)\cos x + g(y)\sin x,$$

then

$$\begin{aligned}
[u(x,y)]_{xx} &= [f(y)\cos x + g(y)\sin x]_{xx} \\
&= f(y)[\cos x]_{xx} + g(y)[\sin x]_{xx} \\
&= -f(y)\cos x - g(y)\sin x,
\end{aligned}$$

and hence

$$[u(x,y)]_{xx} + u(x,y) = 0.$$

Finally,

$$u_{xx} + u = 0 \iff u(x,y) = f(y)\cos x + g(y)\sin x,$$

where f, g are twice continuously differentiable functions on $\mathbf{R}$. $\qquad\square$

Verification with $W|\alpha$: the query

```
D[ u(x,y), x,x ] + u(x,y)=0
```

produces

$$u(x,y) = c_2(y)\sin(x) + c_1(y)\cos(x).$$

Example 1.4.3. Find the general solution of the PDE

$$u_{xy} = 0$$

on $\mathbf{R}^2$.

Solution. As

$$u_{xy} = 0 \iff [u_x]_y = 0,$$

for any twice continuously differentiable function on $\mathbf{R}^2$, we deduce from Example 1.4.1 that

$$u_x = \varphi(x) \iff [u(x,y)]_x = \varphi(x),$$

where φ is a continuously differentiable function.

As

$$[u(x,y)]_x = \varphi(x), \quad \begin{pmatrix} x \\ y \end{pmatrix} \in \mathbf{R}^2$$

is true for all y, it follows that

$$[u(x,y_0)]_x = \varphi(x), \quad (x \in \mathbf{R}),$$

where y_0 is any element of $\mathbf{R}$. The function $h(x) = u(x,y_0)$ is a function (only) in x, and then

$$[h(x)]_x = h'(x) = \varphi(x).$$

Integrating both parts, we get

$$h(x) = \underbrace{\int \varphi(x)\, dx}_{f(x)} + C,$$

where f is a twice continuously differentiable function and C is a constant of integration. More precisely (for each y_0 its own constant!), we write

$$u(x,y_0) = f(x) + g(y_0)$$

As y_0 was arbitrary, we obtain that

$$u_{xy} = 0 \Rightarrow u(x,y) = f(x) + g(y),$$

where f, g are twice continuously differentiable functions.
    The converse is also true:

$$u(x,y) = f(x) + g(y) \Rightarrow [f(x) + g(y)]_{xy}$$
$$= [f'(x)]_y = 0$$
$$\Rightarrow u_{xy} = 0.$$

Therefore,

$$u_{xy} \iff u(x,y) = f(x) + g(y),$$

where f, g are twice continuously differentiable functions on $\mathbf{R}$. $\qquad\square$

    Solution with $W|\alpha$: the query

```
D[ u(x,y), x,y ]=0
```

produces

$$u(x, y) = c_1(x) + c_2(y).$$

Example 1.4.4. Find the general solution of the PDE

$$xu_x = u$$

on $\mathbf{R}^2$.

Solution. We have that

$$x[u(x, y)]_x = u(x, y) \tag{1.4.2}$$

for all $x, y \in \mathbf{R}$. Take some $y_0 \in \mathbf{R}$ for y and substitute in (1.4.2):

$$x[u(x, y_0)]_x = u(x, y_0);$$

then

$$x[h(x)]_x = h(x), \quad (x \in \mathbf{R}),$$

if we write $h(x)$ for $u(x, y_0)$.

We therefore see that $h(x)$ is a solution of the ODE

$$xz'(x) = z(x) \iff z(x) = Cx$$

whose general solution we have obtained in Example 1.3.2.

It follows that

$$h(x) = u(x, y_0) = f(y_0)x$$

(for each y_0 its own constant!). As y_0 was arbitrary, we then get that

$$xu_x = u \Rightarrow u(x, y) = f(y)x.$$

The converse is also true:

$$u(x, y) = f(y)x \Rightarrow x[u(x, y)]_x = x[f(y)x]_x$$
$$= xf(y) = u(x, y)$$
$$\Rightarrow xu_x = u.$$

Therefore,

$$xu_x = u \iff u(x, y) = xf(y),$$

where $f \in C^1(\mathbf{R})$ is a continuously differentiable function, is the general solution of our equation. $\square$

$W|\alpha$: the query

```
x*D[ u(x,y), x ]=u(x,y)
```

brings us

$$u(x, y) = xc_1(y).$$

Note the symbol $*$ for the multiplication; if omitted, the system may not recognize the input as a PDE.

We conclude the chapter with a generalization of the result we have obtained in Example 1.4.1, which we shall need in the chapters that follow.

Proposition 1.4.1. *Let Ω be an x-simple domain in $\mathbf{R}^2$. Then the general solution of the PDE*

$$u_x = 0$$

on Ω is of the form

$$u(x, y) = f(y),$$

where f is a continuously differentiable function on an appropriate open interval in $\mathbf{R}$.

As the reader may recall from calculus, a domain Ω in $\mathbf{R}^2$ is called *x-simple* if the intersection of Ω with any line in $\mathbf{R}^2$ which is parallel to the *x-axis* is either empty, or an interval of this line. A *y-simple* domain in $\mathbf{R}^2$ is defined in an analogous way. Note that any convex domain in $\mathbf{R}^2$ is both x- and y-simple. Due to that, an x-simple domain in $\mathbf{R}^2$ is sometimes called *horizontally convex*, and a y-simple domain in $\mathbf{R}^2$ *vertically convex*.

Proof of Proposition 1.4.1. Let a function $u = u(x, y) \in C^1(\Omega)$ be a solution of the equation

$$u_x = 0$$

on Ω. Suppose that $y_0 \in \mathbf{R}$ is such that the line $y = y_0$, which is parallel to the *x*-axis, meets Ω. Then there exists an open interval I_{y_0}, for which we have that

$$\Omega \cap \{y = y_0\} = I_{y_0} \times \{y_0\} = \left\{ \begin{pmatrix} x \\ y_0 \end{pmatrix} : x \in I_{y_0} \right\}.$$

Then since

$$[u(x, y_0)]_x = 0$$

for all $x \in I_{y_0}$, we get that the function

$$x \mapsto u(x, y_0)$$

is constant on I_{y_0}.

Next, we explain how to determine an open interval mentioned in the formulation of the proposition. Write C for the set of real numbers

$$\{c \in \mathbf{R} : \Omega \cap \{y = c\} \neq \varnothing\}$$

and set

$$\alpha = \inf(C) \quad \text{and} \quad \beta = \sup(C), \tag{1.4.3}$$

where both the infimum and the supremum are taken over the extended real line $\mathbf{R} \cup \{\pm\infty\}$.

Now if a real number c is in the interval

$$I = (\alpha, \beta),$$

then, by (1.4.3), there exist $c_1, c_2 \in C$ satisfying $c_1 < c < c_2$. Hence we have that $c \in C$, for otherwise

$$\Omega = \left\{ \begin{pmatrix} x \\ y \end{pmatrix} \in \Omega : y < c \right\} \cup \left\{ \begin{pmatrix} x \\ y \end{pmatrix} \in \Omega : y > c \right\}$$

is a partition of Ω into a disjoint union of nonempty open subsets, which is impossible due to connectedness of Ω.

We can therefore conclude that

$$u(x, y) = f(y),$$

where f is a continuously differentiable function on I. $\qquad\square$

Corollary 1.4.2. *Consider the PDE*

$$u_x = 0$$

on an open set Ω in $\mathbf{R}^2$ and let $P = \begin{pmatrix} x_0 \\ y_0 \end{pmatrix}$ be any point of Ω. Then we can find an open disk D_P centered at P, which is contained in Ω, such that for every solution $u = u(x, y)$ of the PDE on Ω, there exists a continuously differentiable function f_P on an open interval in $\mathbf{R}$ satisfying

$$u(x, y) = f_P(y)$$

for all points $\begin{pmatrix} x \\ y \end{pmatrix} \in D_P$.

Proof. Existence of a disk D_P is evidently due to the fact that Ω is open.

Consider a function $u \in C^1(\Omega)$ such that

$$u_x = 0$$

everywhere Ω and let v be the restriction of u on D_P. Then

$$v_x = 0$$

everywhere on D_P, and hence, since D_P is x-simple, we have by Proposition 1.4.1 that

$$v(x,y) = //\quad u(x,y)\quad // = f_P(y)$$

for all $\begin{pmatrix} x \\ y \end{pmatrix} \in D_P$, for a suitable function f_P on an open interval in **R**. □

Exercises

(*Use Maple, or W|a, or any other math software of your choice to verify your solutions.*)

1. Find the partial derivatives u_x, u_y, u_{xy} of the following functions u:

$$u(x,y) = \frac{x}{y^2}, \quad u(x,y) = x\sin(x+y),$$

$$u(x,y) = \tan\frac{x^2}{y}.$$

2. Let f be a differentiable function in two variables and let

$$u(x,y) = f\left(x, \frac{x}{y}\right).$$

Find the partial derivatives u_x, u_y, u_{xy}.

3. Show that the function

$$v(x,y) = \ln(x^2 + xy + y^2)$$

is a solution of the PDE

$$xu_x + yu_y = 2$$

on $\Omega = \operatorname{dom}(v)$.

4. Let f be a continuously differentiable function on **R**. Show that the function

$$v(x,y) = f(x^2 + y^2),$$

is a solution of the PDE

$$yu_x - xu_y = 0$$

on $\Omega = \mathbf{R}^2$.

5. Show that the function

$$v(x, y) = \sin(nx)\sinh(ny),$$

where $n \in \mathbf{R}$ is a constant, is a solution of Laplace's equation $u_{xx} + u_{yy} = 0$ on $\Omega = \mathbf{R}^2$.
Recall that the *hyperbolic sine* is the function

$$\sinh(x) = \frac{e^x - e^{-x}}{2},$$

and the *hyperbolic cosine* is the function

$$\cosh(x) = \frac{e^x + e^{-x}}{2}.$$

Before moving on to the problem, verify that

$$\sinh'(x) = \cosh(x) \quad \text{and} \quad \cosh'(x) = \sinh(x).$$

6. For each of the following equations: find order and determine whether it is non-linear, or inhomogeneous linear, or homogeneous linear equation:

(a) $u_t - u_{xx} + xu = 0,$

(b) $u_{tt} - u_{xx} + \sin(x^2) = 0,$

(c) $u_x(1 + u_x^2)^{1/2} + u_y(1 + u_y^3)^{1/3} = 0,$

(d) $u_{tt} + u_{xx} + \sqrt{1 + u} = 0.$

Justify your answers!

7. Find the general solution of each of the following ODEs (first, on any suitable open interval in $\mathbf{R}$, and then, if possible, on the whole of $\mathbf{R}$):

(a) $2xy' = y$; (b) $x^2y' = y.$

8. Obtain the general solution of the PDE

$$u_{xy} - u_x = 0$$

on $\mathbf{R}^2$ by reducing it to an appropriate ODE.

9. Let Ω be the complement of the "nonnegative part" of the x-axis to $\mathbf{R}^2$,

$$\Omega = \mathbf{R}^2 \setminus \left\{ \begin{pmatrix} x \\ 0 \end{pmatrix} : x \geq 0 \right\}$$

$$= \left\{ \begin{pmatrix} x \\ y \end{pmatrix} \in \mathbf{R}^2 : x < 0, \text{ or } y \neq 0 \right\}.$$

Show that Ω is a domain in $\mathbf{R}^2$ (i. e., that Ω is both open and connected), and then justify the following claims from [2, p. 121]:

(i) the function

$$v(x, y) = \begin{cases} x^3 & \text{if } x, y > 0, \\ 0 & \text{otherwise} \end{cases}$$

is continuously differentiable on Ω, and $v_y \equiv 0$ identically on Ω;

(ii) the general solution of the PDE

$$u_y = 0$$

on Ω *cannot* be written in the form

$$u(x, y) = f(x),$$

where f is a continuously differentiable function in one variable, since no such a function exists for the function v above (cf. Proposition 1.4.1; observe also that Ω is *not* y-simple).

(iii) Verify whether or not the function

$$w(x, y) = \begin{cases} x^2 & \text{if } x, y > 0, \\ 0 & \text{otherwise} \end{cases}$$

is continuously differentiable on Ω, and if yes, explain whether or not *it* can be used to prove (ii).

2 Change of variables in PDEs

The material in the *following* chapters requires a thorough understanding of the technique of change of variables in a PDE. If so, the reader can skim through the material in this chapter in order to learn the corresponding terminology and the results. Otherwise, the material should be studied with considerable care. Furthermore, checking, before returning to reading of this chapter, whether the material in Section A.1 of the Appendix to this book is familiar, would be a good idea.

2.1 Change of variables in a PDE: "direct" approach

Basically, as the reader may remember, there are two approaches to changing of variables in PDEs (that could be called "direct" and "indirect," respectively).

Example 2.1.1 (Direct approach to change of variables). Find the general solution of the PDE

$$xu_x + yu_y = u$$

on the domain

$$\Omega = \left\{ \begin{pmatrix} x \\ y \end{pmatrix} \in \mathbf{R}^2 : x > 0 \right\} = \mathbf{R}^+ \times \mathbf{R}$$

in $\mathbf{R}^2$ by performing the change of variables from x, y to new variables s, t where

$$\begin{cases} s = x, \\ t = \dfrac{y}{x}. \end{cases}$$

Solution. As the reader may know, the first and foremost thing to do is to check whether the change of variables is invertible (roughly speaking, it means that it is possible to express x, y via s, t in a *unique* way) and then to check that the variables s, t (resp. x, y) viewed as *functions* in the variables x, y (resp., s, t) are continuously differentiable. In a word, the change of variables must be C^1-invertible.

It is indeed the case in our example:

$$\begin{cases} s = x, \\ t = \dfrac{y}{x} \end{cases} \iff \begin{cases} x = s, \\ y = st \end{cases}$$

for every point $\begin{pmatrix} x \\ y \end{pmatrix} \in \Omega$.

https://doi.org/10.1515/9783110677256-002

Before going on, we would like to remind the reader (as we did above) that the subscript x in the notation $u_x(x,y)$ is "merely a reference to the first argument of the function $u(x,y)$."

To proceed with the change of variables: starting with the equation

$$xu_x(x,y) + yu_y(x,y) = u(x,y),$$

we simply replace x with s and y with st:

$$s\underbrace{u_x(s,st)}_{D_1} + st\underbrace{u_y(s,st)}_{D_2} = \underbrace{u(s,st)}_{w}, \tag{2.1.1}$$

or

$$sD_1 + stD_2 = w$$

for short, where D_1, D_2, and w are defined above.

The next step is, naturally,

to rewrite (2.1.1) as a PDE in variables s, t with respect the (new) unknown function w.

This goal is achievable, once we apply the chain rule to the function

$$w(s,t) = u(s,st).$$

We differentiate the function w by s with the use of the chain rule:

$$w_s = [u(s,st)]_s = u_x(s,st)[s]_s + u_y(s,st)[st]_s$$

whence

$$w_s = D_1 + D_2 t.$$

Similarly, differentiation of w by t gives us

$$w_t = [u(s,st)]_t = u_x(s,st)[s]_t + u_y(s,st)[st]_t,$$

whence

$$w_t = D_2 s$$

Thus we obtain the system

$$\begin{cases} w_s = D_1 + D_2 t, \\ w_t = D_2 s, \end{cases}$$

whose solution is evidently

$$D_2 = \frac{w_t}{s} \quad \text{and} \quad D_1 = w_s - \frac{tw_t}{s}$$

(recall that, by the conditions, $s = x$ is nonzero everywhere on Ω).

Now we substitute the expressions for D_1 and D_2 into the equation

$$sD_1 + stD_2 = w$$

(which is a short form of the equation (2.1.1)), thereby getting that

$$s\left(w_s - \frac{tw_t}{s}\right) + st\frac{w_t}{s} = w,$$

or

$$sw_s = w. \tag{2.1.2}$$

We have actually solved this equation in Example 1.4.4 from Chapter 1:

$$sw_s = w \iff w = w(s,t) = sf(t).$$

We conclude therefore that

$$u(x,y) = /\!/ \quad w(s,t) = sf(t) \quad /\!/ = xf\left(\frac{y}{x}\right),$$

where f is a continuously differentiable function.

Thus any solution of the original equation

$$xu_x + yu_y = u$$

is of the form

$$u(x,y) = xf\left(\frac{y}{x}\right), \tag{2.1.3}$$

where f is a continuously differentiable function.

Conversely, it is easy to check that any function of the above form satisfies our equation (in fact, there is no need for that, since we knew the general solution of the PDE (2.1.2)). Hence (2.1.3) is the general solution of our equation, and we are done. $\square$

The query to $W\!\vert\alpha$ to verify our solution can be

```
x*D[ u[x,y], x ] + y*D[ u[x,y], y ] = u[x,y]
```

and it produces the following answer:

$$u(x, y) = xc_1\left(\frac{y}{x}\right).$$

Note the square brackets enclosing the variables in

```
u[x,y]
```

in the query; it is also admissible, and in fact, it is safer in some cases, for it is the standard way to handle functions in the Wolfram language.

Now let us see how the direct change of variables, we have performed above, can be done in Maple. The following commands can be entered either one-by-one, or, preferably, as a whole, as a *program* in the Maple language.

```
with(PDEtools):

PDE := x*diff(u(x, y), x)
          +y*diff(u(x, y), y) = 0;

varchange := {x = s, y = s*t};
dchange(varchange, PDE);

simplify( dchange(varchange, PDE) );
```

Let us discuss the program briefly. The first command loads a necessary package (PDEtools), from which we take a required command (dchange) which is not available by default. The second command describes our PDE, and the third command our change of variables. The fourth command performs the change of variables. The fifth, and the final, simplifies the result (we could do without the fourth command, but it is always worth studying the "raw data").

The last line of the output Maple produces, once the program above is entered and executed, is

$$\left(\frac{\partial}{\partial s}\, u(s, t)\right)s = 0$$

which agrees with our result above shown in (2.1.2) (note that the same letter u is used in the original and in the transformed equation, which is an understandable thing as far as a math *software* is concerned; more on that below).

Now that we have seen how the method (the direct approach to change of variables in PDEs) works, it is a good idea to review the main steps in a more general settings.

Consider (for simplicity's sake) a first-order PDE

$$F \begin{pmatrix} x \\ y \\ u(x,y) \\ u_x(x,y) \\ u_y(x,y) \end{pmatrix} = 0, \quad \begin{pmatrix} x \\ y \end{pmatrix} \in \Omega \qquad (2.1.4)$$

on a domain Ω in $\mathbf{R}^2$.

Let

$$\begin{cases} s = S(x,y), \\ t = T(x,y), \end{cases} \quad \Longleftrightarrow \quad \begin{cases} x = X(s,t), \\ y = Y(s,t) \end{cases}$$

be a C^1-invertible change of variables on Ω.

Formally, the fact that the change of variables above is C^1-invertible means that the following functions Φ and Ψ are continuously differentiable and are inverses of one another:

$$\Phi \begin{pmatrix} x \\ y \end{pmatrix} = \begin{pmatrix} S(x,y) \\ T(x,y) \end{pmatrix}, \quad \begin{pmatrix} x \\ y \end{pmatrix} \in \Omega$$

and

$$\Psi \begin{pmatrix} s \\ t \end{pmatrix} = \begin{pmatrix} X(s,t) \\ Y(s,t) \end{pmatrix}, \quad \begin{pmatrix} s \\ t \end{pmatrix} \in \Lambda,$$

where

$$\Lambda = \Phi(\Omega).$$

As Ψ is the inverse of Φ on Ω, we have

$$(\Psi \circ \Phi) \begin{pmatrix} x \\ y \end{pmatrix} = \mathrm{id}_\Omega \begin{pmatrix} x \\ y \end{pmatrix} = \begin{pmatrix} x \\ y \end{pmatrix}$$

for all $\begin{pmatrix} x \\ y \end{pmatrix} \in \Omega$. Since

$$\begin{pmatrix} x \\ y \end{pmatrix} = (\Psi \circ \Phi) \begin{pmatrix} x \\ y \end{pmatrix} = \Psi \begin{pmatrix} S(x,y) \\ T(x,y) \end{pmatrix}$$
$$= \begin{pmatrix} X(S(x,y), T(x,y)) \\ Y(S(x,y), T(x,y)) \end{pmatrix},$$

we obtain that

$$x = X(S(x, y), T(x, y)),$$
$$y = Y(S(x, y), T(x, y)), \tag{2.1.5}$$

for all $\begin{pmatrix} x \\ y \end{pmatrix} \in \Omega$.

Similarly, using the fact that

$$\Phi \circ \Psi = \mathrm{id}_\Lambda,$$

we get

$$s = S(X(s, t), Y(s, t)),$$
$$t = T(X(s, t), Y(s, t)),$$

for all $\begin{pmatrix} s \\ t \end{pmatrix} \in \Lambda$.

Then the *direct approach to the change of variables* $x, y \to s, t$ in the PDE (2.1.4) is as follows:

I. We *substitute* $X(s, t)$ for x and $Y(s, t)$ for y in (2.1.5), thereby getting the equation

$$F\begin{pmatrix} X(s, t) \\ Y(s, t) \\ u(X(s, t), Y(s, t)) \\ u_x(X(s, t), Y(s, t)) \\ u_y(X(s, t), Y(s, t)) \end{pmatrix} = 0, \quad \begin{pmatrix} s \\ t \end{pmatrix} \in \Lambda \tag{2.1.6}$$

in variables s, t on Λ, and then we try to rewrite it equivalently as a *partial differential equation* for the unknown function

$$w = w(s, t) = u(X(s, t), Y(s, t))$$

on Λ. Observe also that, by (2.1.5), we have

$$w(S(x, y), T(x, y))$$
$$= u(\underbrace{X(S(x, y), T(x, y))}_{x}, \underbrace{Y(S(x, y), T(x, y))}_{y})$$

or

$$u(x, y) = w(S(x, y), T(s, y)) \tag{2.1.7}$$

for all $\begin{pmatrix} x \\ y \end{pmatrix} \in \Omega$. More elegantly put, since

$$w = u \circ \Psi,$$

then

$$u = w \circ \Psi^{-1},$$

or

$$u = w \circ \Phi,$$

which amounts to the same thing as (2.1.7).

Importantly, by the construction,

> a function $u = u(x, y) \in C^1(\Omega)$ is a solution of the equation (2.1.4)
> on Ω if and only if the function $w(s, t) = u(X(s, t), Y(s, t)) \in C^1(\Lambda)$ (2.1.8)
> is a solution of the equation (2.1.6) on Λ,

since, to emphasize, $x, y \to s, t$ is a C^1-invertible change of variables.

II. However, in order to rewrite equation (2.1.6) as a partial differential equation, we have to express the functions

$$D_1 = D_1(s, t) = u_x(X(s, t), Y(s, t))$$

and

$$D_2 = D_2(s, t) = u_y(X(s, t), Y(s, t))$$

in terms of w, w_s, w_t.

We solve this problem by applying the chain rule. Now

$$[u(X(s, t), Y(s, t))]_s$$
$$= u_x(X(s, t), Y(s, t))X_s + u_y(X(s, t), Y(s, t))Y_s$$

and

$$[u(X(s, t), Y(s, t))]_t$$
$$= u_x(X(s, t), Y(s, t))X_s + u_y(X(s, t), Y(s, t))Y_s,$$

where, say X_s is the short form for

$$X_s(s, t) = [X(s, t)]_s = \frac{\partial X(s, t)}{\partial s}.$$

In other words,

$$w_s = D_1 X_s + D_2 Y_s,$$
$$w_t = D_1 X_t + D_2 Y_t$$

whence, by Cramer's rule,

$$D_1 = \frac{\begin{vmatrix} w_s & Y_s \\ w_t & Y_t \end{vmatrix}}{\begin{vmatrix} X_s & Y_s \\ X_t & Y_t \end{vmatrix}} \quad \text{and} \quad D_2 = \frac{\begin{vmatrix} X_s & w_s \\ X_t & w_t \end{vmatrix}}{\begin{vmatrix} X_s & Y_s \\ X_t & Y_t \end{vmatrix}}$$

The function

$$\begin{vmatrix} X_s & Y_s \\ X_t & Y_t \end{vmatrix} = \det \frac{\partial(X, Y)}{\partial(s, t)}$$

in both denominators is the Jacobian determinant of the back change $s, t \rightarrow x, y$, and hence, by Proposition A.1.1 from Section A.1 in the Appendix, it is nonzero everywhere on Λ.

To sum up: the substitutions $x \rightarrow X(s, t)$ and $y \rightarrow Y(s, t)$ produce a new PDE

$$G(s, t, w(s, t), w_s(s, t), w_t(s, t)) = 0$$

for the function w depending on s, t on Λ, which is called the *transformed* PDE. In this context, the equation

$$F(x, y, u, u_x, u_y) = 0,$$

with which we have started, is called the *original* equation.

Now suppose that we have somehow obtained the general solution of the transformed PDE:

$$G(s, t, w(s, t), w_s(s, t), w_t(s, t)) = 0$$
$$\Updownarrow$$
$$w(s, t) = g(s, t),$$

where g runs over some class $\mathcal{G}$ of C^1-functions on Λ.

Next, assume that a function $u(x, y) \in C^1(\Omega)$ is a solution of the original PDE on Ω. It is equivalent to the fact that, by (2.1.8), the function

$$w(s, t) = u(X(s, t), Y(s, t))$$

is a solution of the transformed PDE on Λ, which, in turn, means, that

$$w(s, t) = u(X(s, t), Y(s, t)) = g(s, t) \tag{2.1.9}$$

for a suitable $g \in \mathcal{G}$.

As we have discussed above, the substitutions

$$s \to S(x,y), \quad t \to T(x,y)$$

in $X(s,t)$ and in $Y(s,t)$ produce x and y, respectively. It follows that (2.1.9) is equivalent to

$$u(x,y) = g(S(x,y), T(x,y))$$

for a suitable $g \in \mathcal{G}$.

To sum up: a function $u = u(x,y)$ is a solution of the original equation if and only if

$$u(x,y) = g(S(x,y), T(x,y))$$

for an appropriate function $g \in \mathcal{G}$.

We have to note that in practice most people, most of the time, prefer a simplified technique.

We start with the functional equation

$$w(s,t) = u(x,y),$$

which is (of course) a simplified version of the functional equation

$$w(s,t) = u(X(s,t), Y(s,t))$$

above, thereby assuming that x and y depend on s,t in the manner prescribed by the corresponding change of variables. In short, we treat x,y are *functions* in s,t.

So x,y are treated like objects of *dual* nature:

-   on one hand, we (continue to) treat them as *variables*;
-   on the other hand, we treat them as *functions* in s,t.

(much like the dual nature of light, meaning the well-known *wave-particle* duality of light).

OK, we differentiate the relation $w(s,t) = u(x,y)$ first by s, and then by t, by applying the chain rule:

$$w_s = u_x(x,y)x_s + u_y(x,y)y_s,$$
$$w_t = u_x(x,y)x_t + u_y(x,y)y_t.$$

Then, as above, by Cramer's rule,

$$u_x(x,y) = \frac{\begin{vmatrix} w_s & y_s \\ w_t & y_t \end{vmatrix}}{\begin{vmatrix} x_s & y_s \\ x_t & y_t \end{vmatrix}} \quad \text{and} \quad u_y(x,y) = \frac{\begin{vmatrix} x_s & w_s \\ x_t & w_t \end{vmatrix}}{\begin{vmatrix} x_s & y_s \\ x_t & y_t \end{vmatrix}}. \tag{2.1.10}$$

We then replace/change

$$u_x(x,y) \quad \text{and} \quad u_y(x,y)$$

with their expressions in the right-hand sides of (2.1.10), and (of course) replace $u(x,y)$ with $w(s,t)$.

Even simpler,
–   replace u with w;
–   replace u_x, u_y with their expressions from (2.1.10).

Too bad to do so, but every so often the same variable is used both in the original and in the transformed equation (despite a clear evidence to the contrary, for the functions/variables u and w above need *not* be the same). Everywhere below, we will always use a different variable in the transformed PDE.

Once the general solution of the transformed equation is found, we again write

$$u(x,y) = w(s,t) = g(s,t), \quad (g \in \mathcal{G})$$

and replace s, t with their expressions from the change of variables.

Say, in our previous example we had to perform the change of variables

$$\begin{cases} s = x, \\ t = \dfrac{y}{x} \end{cases}$$

in the PDE

$$xu_x + yu_y = u \quad (x \neq 0)$$

on $\Omega = \mathbf{R}^+ \times \mathbf{R}$. We have seen that

$$w(s,t) = sf(t)$$

is the general solution of the transformed equation. To summarize the solution, we can write

$$u(x,y) = // \quad w(s,t) = sf(t) \quad // = xf\left(\frac{y}{x}\right),$$

thereby referring to the general solution of the transformed equation, and obtaining the general solution of the original solution by means of replacing the new variables with the old ones.

2.2 Inverse function theorem

Now let us turn to the discussion of the "indirect" approach to a change of variables.
As it has been stressed above, given a change of variables in the form

$$\begin{cases} s = S(x,y), \\ t = T(x,y) \end{cases} \tag{2.2.1}$$

on an open set $\Omega \subseteq \mathbf{R}^2$, where $S, T \in C^1(\Omega)$, for a PDE

$$F(x, y, u, u_x, u_y, \ldots) = 0,$$

the first thing to do is to verify that the change of variables is C^1-invertible on Ω,

or, equivalently, to establish that the function

$$\Phi\begin{pmatrix} x \\ y \end{pmatrix} = \begin{pmatrix} S(x,y) \\ T(x,y) \end{pmatrix}, \quad \begin{pmatrix} x \\ y \end{pmatrix} \in \Omega$$

has a continuously differentiable inverse.

However, this problem may be very hard indeed, and in this section we shall discuss the best result that exists which provides a *sufficient* condition under which Φ is at least *locally* C^1-invertible. This result is called the *Inverse Function Theorem,* or the *IFT*:

Theorem (Inverse function theorem). *Let $\Phi : \Omega \to \mathbf{R}^2$, where*

$$\Phi\begin{pmatrix} x \\ y \end{pmatrix} = \begin{pmatrix} f(x,y) \\ g(x,y) \end{pmatrix}$$

be a continuously differentiable map on an open set $\Omega \subseteq \mathbf{R}^2$ and let $P = \begin{pmatrix} x_0 \\ y_0 \end{pmatrix}$ be a point of Ω. Then the following are equivalent:
  (i) *the Jacobian determinant*

$$\det J_\Phi(x_0, y_0) = \det \frac{\partial(f,g)}{\partial(x,y)}(x_0, y_0)$$

$$= \begin{vmatrix} f_x(x_0, y_0) & f_y(x_0, y_0) \\ g_x(x_0, y_0) & g_y(x_0, y_0) \end{vmatrix} \neq 0$$

is nonzero at $P = \begin{pmatrix} x_0 \\ y_0 \end{pmatrix} \in \Omega$;
  (ii) *Φ is locally C^1-invertible near the point P, that is, there exists an open neighborhood $O_P \subseteq \Omega$ of P such that the restriction Ψ of Φ on O_P is invertible and Ψ^{-1} is continuously differentiable;*

(iii) *the change of variables*

$$\begin{cases} s = f(x, y), \\ t = g(x, y) \end{cases}$$

is C^1-invertible in a suitable open neighborhood O_P of P.

We recommend the reader to study the formulation and the proof of the general case of the Inverse Function Theorem (for arbitrary dimensions; an excellent account can be found in [7, Theorem 9.24]); it is especially important if the reader is serious about their mathematics, whatever their chosen field of study.

Example 2.2.1. There is a continuously differentiable map $\Phi : \mathbf{R}^2 \to \mathbf{R}^2$ which is *locally C^1-invertible* near every point of $\mathbf{R}^2$, but which is not C^1-invertible on $\mathbf{R}^2$ (i. e., it does not have the *global* inverse).

Solution. A classical example is constructed over the complex exponential function $f(z) = e^z$ which is periodic, since

$$f(z + 2\pi i) = f(z), \quad (z \in \mathbf{C}),$$

and for which we have that

$$f(x + iy) = e^{x+iy} = e^x(\cos y + i \sin y),$$

where $x, y \in \mathbf{R}$. The corresponding map from $\mathbf{R}^2$ into $\mathbf{R}^2$ is therefore

$$\Phi\begin{pmatrix} x \\ y \end{pmatrix} = \begin{pmatrix} \operatorname{Re} f(x + iy) \\ \operatorname{Im} f(x + iy) \end{pmatrix} = \begin{pmatrix} e^x \cos y \\ e^x \sin y \end{pmatrix},$$

where $\begin{pmatrix} x \\ y \end{pmatrix}$ runs over $\mathbf{R}^2$. We have that

$$J_\Phi(x, y) = \begin{pmatrix} [e^x \cos y]_x & [e^x \cos y]_y \\ [e^x \sin y]_x & [e^x \sin y]_y \end{pmatrix}$$

$$= \begin{pmatrix} e^x \cos y & -e^x \sin y \\ e^x \sin y & e^x \cos y \end{pmatrix},$$

whence

$$\det J_\Phi(x, y) = e^{2x} \cos^2 y + e^{2x} \sin^2 y = e^{2x} \neq 0$$

everywhere on $\mathbf{R}^2$. Therefore, according to the Inverse Function Theorem, Φ is C^1-locally invertible everywhere on $\mathbf{R}^2$.

But Φ is *not* invertible (since the complex exponential function is not). Indeed, a given function is invertible iff it is 1-1 and onto. But Φ is definitely not 1-1 (i. e., there is a pair of distinct points in the domain whose images under Φ are equal), because, for instance,

$$\begin{pmatrix} 1 \\ 0 \end{pmatrix} = \Phi \begin{pmatrix} 0 \\ 0 \end{pmatrix} = \Phi \begin{pmatrix} 0 \\ 2\pi \end{pmatrix} = \begin{pmatrix} 1 \\ 0 \end{pmatrix}. \qquad \qquad \square$$

2.3 Change of variables in a PDE: "indirect" approach

Well, the direct approach to a change of variables in a PDE always works (the result is *guaranteed*... if no mistakes are made...)

...and computers (or, to be precise, programmers) like it very much, since it is easily implemented, but...

...there is another approach to change of variables $x, y \rightarrow s, t$, which is in fact used more commonly...

...this approach is sometimes called the *indirect* approach to change of variables. The idea is as follows:

OK, even if we make it sure that our change of variables is at least locally C^1-invertible, we may be simply *unable* to express x and y explicitly via s and t for the direct approach to work, that is, again, we may be unable to find functions X, Y such that

$$\begin{cases} s = S(x, y), \\ t = T(x, y) \end{cases} \qquad \Longleftrightarrow \qquad \begin{cases} x = X(s, t), \\ y = Y(s, t) \end{cases} \qquad (2.3.1)$$

everywhere on our domain Ω.

Still, something can be done. Let us assume for the moment that we work with such domain Ω that the change of variables (2.3.1) *is* C^1-invertible, that is, both the functions X, Y are continuously differentiable on Ω.

If so, it is possible to start with the familiar relation

$$u(x, y) = w(s, t),$$

but, this time,

it will be the *new variables* s, t that will be treated as objects of dual nature: both as *variables* and as *functions* in x, y.

Now if in the relation

$$u(x, y) = w(s, t)$$

s, t have dual nature, we can differentiate the relation first by x, and then by y:

$$u_x(x, y) = w_s(s, t)s_x + w_t(s, t)t_x$$

and

$$u_y(x, y) = w_s(s, t)s_y + w_t(s, t)t_y,$$

or

$$u_x = w_s s_x + w_t t_x, \qquad (2.3.2)$$
$$u_y = w_s s_y + w_t t_y$$

for short. We then
- replace u with w;
- replace u_x and u_y with their expressions in (2.3.2)

in the original equation...

...in a hope that all occurrences of x and y in the original equation will disappear, and we shall get a PDE for $w = w(s, t)$, the familiar *transformed* equation (in general, however, the result, as to be expected, is *not* guaranteed, unlike in the situation when one uses the direct approach).

Let us consider an example to make things clear.

Example 2.3.1 (Change of variables: "indirect" approach). Use the change of variables

$$\begin{cases} s = xy, \\ t = x/y \end{cases}$$

to solve the PDE

$$xu_x - yu_y = 2x^2.$$

Now what about an open set to consider the equation on? A reasonable answer to this question could be as follows:

on any unspecified domain $\Omega \subseteq \mathbf{R}^2$ on which the change of variables is C^1-invertible.

Observe that for any such domain we must have that

$$y \neq 0,$$

since there is division by y in the change of variables.

Solution of Example 2.3.1. **Step 1.** Find the Jacobian matrix and the Jacobian determinant the change.

As

$$
\begin{cases}
s = xy, \\
t = \dfrac{x}{y},
\end{cases}
$$

we have that

$$
J_\Phi(x,y) = \frac{\partial(s,t)}{\partial(x,y)} = \begin{pmatrix} s_x & s_y \\ t_x & t_y \end{pmatrix}
$$

$$
= \begin{pmatrix} [xy]_x & [xy]_y \\ \left[\dfrac{x}{y}\right]_x & \left[\dfrac{x}{y}\right]_y \end{pmatrix} = \begin{pmatrix} y & x \\ \dfrac{1}{y} & -\dfrac{x}{y^2} \end{pmatrix},
$$

whence

$$
\det J_\Phi(x,y) = -2\frac{x}{y}.
$$

Thus our domain Ω must be such that

$$
\frac{x}{y} \neq 0 \iff t \neq 0
$$

everywhere on Ω. So not only our domain must not meet the line $y = 0$, but also it must not meet the line $x = 0$.

Step 2. Write the basic relation

$$
u(x,y) = w(s,y)
$$

and differentiate it first by x, and then by y, using the chain rule:

$$
u_x = w_s s_x + w_t t_x,
$$
$$
u_y = w_s s_y + w_t t_y,
$$

or

$$
u_x = w_s [xy]_x + w_t \left[\frac{x}{y}\right]_x,
$$
$$
u_y = w_s [xy]_y + w_t \left[\frac{x}{y}\right]_y,
$$

or

$$
u_x = w_s y + w_t \frac{1}{y}, \tag{2.3.3}
$$
$$
u_y = w_s x - w_t \frac{x}{y^2}
$$

(the partial derivatives, we have obtained when calculating the Jacobian matrix, come handy, as the reader can see).

Step 3. Substitute w for u and substitute the expressions for u_x and u_y from (2.3.3) into the original PDE.

OK, since

$$xu_x - yu_y = 2x^2$$

then

$$x\left(w_s y + w_t \frac{1}{y}\right) - y\left(w_s x - w_t \frac{x}{y^2}\right) = 2x^2,$$

or

$$2\frac{x}{y}w_t = 2x^2. \qquad (2.3.4)$$

Step 4. Eliminate, if possible, all occurrences of x and y in the equation obtained at the previous step.

Recall that

$$s = xy \quad \text{and} \quad t = x/y$$

and hence, luckily, what has been left of x and y, is easily expressible via s and t: we obtain from (2.3.4) that

$$tw_t(s,t) = st,$$

or

$$w_t(s,t) = s,$$

since $t \neq 0$.

Thus the original PDE is transformed to the PDE

$$w_t = s \qquad (2.3.5)$$

(or, in other words, it is the *transformed* equation).

Clearly, in turn, the transformed equation is reducible to an ODE (i. e., to a differential equation involving just *one* independent variable).

Indeed, we simply *integrate* both parts of (2.3.5), treating s as a constant:

$$[w(s,t)]_t = s \Rightarrow w(s,t) = ts + f(s),$$

and hence

$$u(x, y) = /\!/ \quad w(s, t) = ts + f(s) \quad /\!/$$
$$= x^2 + f(xy),$$

where f is a continuously differentiable function. Conversely, we easily check that $u(x, y)$ satisfies the original equation.

Therefore,

$$u(x, y) = x^2 + f(xy),$$

where f is a continuously differentiable function, is the general solution of our equation on any domain Ω, on which our change of variables is C^1-invertible. □

As we have discussed above, the computers (would) "prefer" the direct approach to change of variables. However, the indirect approach can be also modeled with math software. Let us demonstrate how it is done with Maple. A couple of words beforehand:

- strings that start with a > symbol are Maple commands; strings that follow are the Maple responses;
- strings that start with a # symbol are comments;
- the first command

    restart;

  restarts the system anew, and, in effect, all objects introduced earlier are removed (the command can be useful if one switches from one problem to another, and wishes to make sure that there is no conflict of variables, etc.);
- the second and the third commands show how one can introduce two-variable functions (to introduce a function in more than two variables, one uses a similar command); a function in one variable is introduced with a command like

    f := x -> 2*x^2-3*x+4;

Now the promised solution with Maple:

```
>  restart;
>  s := (x,y) -> x*y; # think of s as a FUNCTION
```

$$s := (x, y) \rightarrow xy$$

```
>  t := (x,y) -> x/y; # think of t as a FUNCTION
```

$$t := (x, y) \rightarrow \frac{x}{y}$$

```
>  u_x:= w_s*diff(s(x,y),x)+w_t * diff(t(x,y),x);
```

$$u\_x := w\_s\, y + \frac{w\_t}{y}$$

> u_y:= w_s * diff(s(x,y),y)+ w_t * diff(t(x,y),y);

$$u\_y := w\_s\, x - \frac{w\_t\, x}{y^2}$$

> # Here u_x, u_y, w_s, w_t are just VARIABLES for Maple; they keep the expressions for the corresponding derivatives;

> # Substitute what we introduced above in the original equation:

> x*u_x - y*u_y - 2*x^2;

$$x\left(w\_s\, y + \frac{w\_t}{y}\right) - y\left(w\_s\, x - \frac{w\_t\, x}{y^2}\right) - 2x^2$$

> simplify(x*u_x - y*u_y - 2*x^2); # simplification of the previous output

$$2\,\frac{x\,(w\_t - x\,y)}{y}$$

Now we can finish the solution ourselves, by replacing x/y and xy with their expressions via s and t, thereby getting the PDE

$$2t(w_t - s) = 0,$$

which is equivalent to the PDE

$$w_t = s,$$

on the domain we work with, which is exactly what we have had above.

Exercises

1. (*Direct approach*). Perform the change of variables

$$\begin{cases} x = s\cos t, \\ y = s\sin t \end{cases}$$

(observe that s, t are in fact the polar coordinates) in the linear PDE

$$yu_x - xu_y = 0$$

on the domain

$$\Omega = \left\{ \begin{pmatrix} x \\ y \end{pmatrix} : x > 0 \right\} = \mathbf{R}^+ \times \mathbf{R}$$

in $\mathbf{R}^2$ to obtain the general solution of the PDE on Ω.

   2. (*Indirect approach*). Perform the change of variables

$$\begin{cases} s = \ln(x), \\ t = \ln(y + \sqrt{1 + y^2}) \end{cases}$$

to find the general solution of the linear PDE

$$xu_x + \sqrt{1 + y^2}u_y = 0$$

(on any unspecified domain Ω in $\mathbf{R}^2$ on which the change of variables is C^1-invertible; use math software to solve the transformed equation, or, alternatively, read the proof of and then apply Proposition 3.1.1 from the next chapter).

3 First-order linear equations

The goal of this chapter is to study how to obtain the general solution of a linear PDE

$$au_x + bu_y + cu = d(x, y), \tag{3.1.1}$$

where $a, b, c \in \mathbf{R}$ and either a, or b is nonzero, and $d(x, y)$ is a continuous function. We first explain how to solve the equation

$$au_x + bu_y = 0,$$

then the equation

$$au_x + bu_y + cu = 0,$$

and, finally, the equation (3.1.1).

Proposition 3.1.1. *Let a, b be real numbers satisfying $a^2 + b^2 \neq 0$. Then the change of variables*

$$\begin{cases} s = ax + by, \\ t = bx - ay \end{cases}$$

transforms the linear PDE

$$au_x + bu_y = 0$$

on $\Omega = \mathbf{R}^2$ to the PDE

$$(a^2 + b^2)w_s = 0 \iff w = w(s, t) = f(t).$$

Consequently, the general solution of the PDE

$$au_x + bu_y = 0$$

on $\Omega = \mathbf{R}^2$ is

$$u(x, y) = /\!/ \quad w(s, t) = f(t) \quad /\!/ = f(bx - ay),$$

where f is a continuously differentiable function on $\mathbf{R}$.

Observe that the graph of any particular solution

$$u(x, y) = f(bx - ay)$$

of the linear PDE

$$au_x + bu_y = 0$$

https://doi.org/10.1515/9783110677256-003

in Proposition 3.1.1 is a rather easily recognizable ruled surface (one which is made up of straight lines). Indeed, the domain $\Omega = \mathbf{R}^2$, on which we consider this PDE, is a union of the (parallel) straight plane lines

$$bx - ay = C, \tag{3.1.2}$$

where $C \in \mathbf{R}$ is an arbitrary constant, and so Ω may be thought of as being swept by a moving line of the form (3.1.2). Moreover, the function $u(x, y) = f(bx - ay)$ is *constant* on any of the straight lines of the form (3.1.2), since

$$bx_0 - ay_0 = bx_1 - ay_1 = C \Rightarrow u(x_0, y_0) = u(x_1, y_1) = f(C)$$

for all $x_0, y_0, x_1, y_1 \in \mathbf{R}$. This means that the graph of the solution $u = u(x, y)$ may be thought of as being swept by a moving straight line of the form

$$\begin{cases} bx - ay = C, \\ z = f(C), \end{cases} \tag{3.1.3}$$

where C runs over $\mathbf{R}$. Note that for every $C \in \mathbf{R}$, the curve (3.1.3) is the translation of the line

$$\begin{cases} bx - ay = C, \\ z = 0 \end{cases}$$

in the direction of the vector $\begin{pmatrix} 0 \\ 0 \\ f(C) \end{pmatrix}$ parallel to the z-axis.

Figure 3.1.1 provides an illustration to the above considerations: a number of the plane lines of the form (3.1.2) are shown in gray, their translations (3.1.3), which lie on the graph of a particular solution $u = u(x, y)$, are shown in black.

Proof of Proposition 3.1.1. First, we have to make sure that the change of variables suggested in the formulation of the proposition is C^1-invertible. It is easy, if one remembers one's linear algebra.

We observe that the change of variables can be written as

$$\begin{pmatrix} s \\ t \end{pmatrix} = T \begin{pmatrix} x \\ y \end{pmatrix},$$

where the matrix

$$T = \begin{pmatrix} a & b \\ b & -a \end{pmatrix}$$

is called, as it must be remembered from linear algebra, the *transition matrix* of the change of variables (in fact, it is also the *Jacobian matrix* of our change of variables).

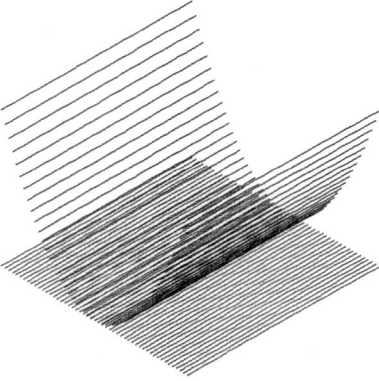

Figure 3.1.1: Plane lines $bx - ay = C$ and their translations.

The matrix T is invertible: indeed,

$$T^2 = \begin{pmatrix} a & b \\ b & -a \end{pmatrix} \begin{pmatrix} a & b \\ b & -a \end{pmatrix}$$

$$= \begin{pmatrix} a^2 + b^2 & 0 \\ 0 & a^2 + b^2 \end{pmatrix}$$

$$= (a^2 + b^2)I,$$

where I is the identity matrix, and then

$$T^{-1} = \frac{1}{a^2 + b^2} T = \frac{1}{a^2 + b^2} \begin{pmatrix} a & b \\ b & -a \end{pmatrix}.$$

Thus

$$\begin{pmatrix} s \\ t \end{pmatrix} = T \begin{pmatrix} x \\ y \end{pmatrix} \Rightarrow \begin{pmatrix} x \\ y \end{pmatrix} = T^{-1} \begin{pmatrix} s \\ t \end{pmatrix},$$

whence

$$\begin{cases} x = \dfrac{as + bt}{a^2 + b^2}, \\ y = \dfrac{bs - at}{a^2 + b^2}. \end{cases}$$

We therefore obtain that

$$\begin{cases} s = ax + by, \\ t = bx - ay, \end{cases} \iff \begin{cases} x = \dfrac{as + bt}{a^2 + b^2}, \\ y = \dfrac{bs - at}{a^2 + b^2}, \end{cases}$$

and so we are dealing with a C^1-invertible change of variables.

Now we have a choice to make: either we apply the direct, or the indirect approach to change of variables. The indirect approach is somewhat easier in our case.

So we start with the relation

$$u(x, y) = w(s, t)$$

and differentiate it first by x, and then by y:

$$u_x = w_s s_x + w_t t_x,$$
$$u_y = w_s s_y + w_t t_y,$$

or ($s = ax + by$, $t = bx - ay$),

$$u_x = aw_s + bw_t, \tag{3.1.4}$$
$$u_y = bw_s - aw_t.$$

Now, by (3.1.4),

$$au_x + bu_y = a(aw_s + bw_t) + b(bw_s - aw_t)$$
$$= (a^2 + b^2)w_s,$$

and our PDE is indeed transformed to the equation

$$(a^2 + b^2)w_s = 0 \iff w_s = 0$$
$$\iff w = w(s, t) = f(t),$$

where, to justify the first equivalence, we have used the condition $a^2 + b^2 \neq 0$.

Finally,

$$u(x, y) = /\!/ \quad w(s, t) = f(t) \quad /\!/ = f(bx - ay),$$

where f is a continuously differentiable function on $\mathbf{R}$.

Conversely, it is easy to verify that any function of the above form is a solution of the original PDE. □

Solution with $W|a$: the query

```
a*D[ u(x,y),x ] + b*D[ u(x,y),y ]=0
```

produces

$$u(x, y) = c_1\left(y - \frac{bx}{a}\right).$$

This answer is different from that one of ours, and the latter is more sound (since, of course, a could be equal to zero). But if $a \neq 0$, then any function of the form

$$f\left(y - \frac{b}{a}x\right)$$

can be rewritten a function of the form

$$g(bx - ay),$$

and vice versa.

Example 3.1.1. (i) Find the general solution of the PDE

$$4u_x - 7u_y = 0;$$

(ii) find the solution of this PDE satisfying the (initial) condition

$$u(0, y) = \sin y$$

for all $y \in \mathbf{R}$.

What we have above is a typical problem of the theory of PDEs: first find the general solution (if possible, of course), and then find the function which satisfies a certain initial condition (a setup like that is called a *Cauchy problem*; sometimes, however, we may skip the first step, if the general solution is too hard to come by).

Keeping in mind that we deal with the two-variable case, a typical Cauchy problem will require a solution $u(x, y)$ take prescribed values on a given curve Γ in $\mathbf{R}^2$ (in the plane Oxy):

$$u(x, y) = \varphi(x, y), \quad \begin{pmatrix} x \\ y \end{pmatrix} \in \Gamma.$$

In our case, the (initial) curve Γ is the straight line

$$x = 0$$

in $\mathbf{R}^2$.

Solution of Example 3.1.1. (i) By Proposition 3.1.1, the general solution of our PDE is given in the form

$$u(x, y) = \|\ \ f(bx - ay), a = 4, b = -7\ \ \|$$
$$= f(-7x - 4y),$$

where $f \in C^1(\mathbf{R})$ is a continuously differentiable function.

(ii) Let us find the solution which satisfies the initial condition above. This solution must be of the form

$$u(x, y) = f(-7x - 4y),$$

whence

$$u(0, y) = f(-4y).$$

Now, for a function we are looking for, we must have

$$u(0, y) = \sin y \iff f(-4y) = \sin y$$

for all $y \in \mathbf{R}$.

Letting z denote $-4y$, we get

$$f(z) = \sin\left(-\frac{z}{4}\right),$$

and hence the function

$$u(x, y) = \sin\frac{7x + 4y}{4}$$

is the solution we were looking for, which can be verified after substitution of this function into the equation and into the initial condition. $\square$

Proposition 3.1.2. *Let*

$$au_x + bu_y + cu = 0 \tag{3.1.5}$$

be a first-order homogeneous linear PDE with constant coefficients $a, b, c \in \mathbf{R}$ such that $a^2 + b^2 \neq 0$. Then:

(i) *the change of variables*

$$\begin{cases} s = ax + by, \\ t = bx - ay \end{cases}$$

transforms the PDE (3.1.5) *to the PDE*

$$(a^2 + b^2)w_s + cw = 0, \tag{3.1.6}$$

which is, in turn, reducible to the ODE

$$z'(s) + \lambda z(s) = 0,$$

where

$$\lambda = \frac{c}{a^2 + b^2};$$

(ii) *the general solution of the transformed equation* (3.1.6) *is*

$$(a^2 + b^2)w_s + cw = 0 \iff w(s,t) = e^{-\lambda s}f(t);$$

(iii) *consequently, the general solution of the PDE* (3.1.5) *on* $\Omega = \mathbf{R}^2$ *is given by*

$$u(x,y) = \mathbin{/\!/} \quad w(s,t) = e^{-\lambda s}f(t) \quad \mathbin{/\!/}$$
$$= e^{-\lambda(ax+by)}f(bx - ay),$$

where f is a continuously differentiable function on **R**.

Proof. (i) Let a function $u = u(x,y)$ be a solution of the PDE (3.1.5) on $\mathbf{R}^2$. According to the proof of Proposition 3.1.1, given that

$$u(x,y) = w(s,t),$$

our change of variables transforms the function

$$au_x + bu_y$$

to the function

$$(a^2 + b^2)w_s.$$

In fact, we can state this as simply as

$$au_x + bu_y = (a^2 + b^2)w_s.$$

But then

$$au_x + bu_y + cu = (a^2 + b^2)w_s + cw.$$

Accordingly, our equation is indeed transformed to the PDE

$$(a^2 + b^2)w_s + cw = 0,$$

which is evidently equivalent to the PDE

$$w_s + \lambda w = 0, \tag{3.1.7}$$

where

$$\lambda = \frac{c}{a^2 + b^2}.$$

(ii) The equation (3.1.7) is reducible to the ODE

$$z'(s) + \lambda z(s) = 0$$

whose general solution is

$$z(s) = Ce^{-\lambda s},$$

where $C \in \mathbf{R}$ is an arbitrary constant.

Therefore,

$$w(s,t) = \underbrace{f(t)}_{C} e^{-\lambda s}.$$

(iii) By recalling the equality

$$u(x,y) = w(s,t),$$

we obtain that u must be of the form

$$u(x,y) = //\quad w(s,t) = e^{-\lambda s}f(t)\quad //$$
$$= e^{-\lambda(ax+by)}f(bx - ay).$$

Conversely, it is easy to check that this function is indeed a solution of the original PDE. □

It is interesting to compare what $W|\alpha$ "thinks" about the equation we have worked with. The query

```
a*D[ u[x,y],x ] + b*D[ u[x,y],y ] + c*u[x,y]=0
```

produces the answer

$$u(x,y) = e^{-(cx)/a}c_1\left(y - \frac{bx}{a}\right)$$

which should be taken with a grain of salt (that might be a good idea any time one uses math software).

Recall one important result from the theory of ODEs.

Example 3.1.2. Given a constant λ and a continuous function $g(s)$, the general solution of the ODE

$$z'(s) + \lambda z(s) = g(s),$$

is

$$z(s) = e^{-\lambda s}\left(\int e^{\lambda s}g(s)\,ds + C\right)$$
$$= \underbrace{e^{-\lambda s}\int e^{\lambda s}g(s)\,ds}_{z_{\text{prt}}} + \underbrace{Ce^{-\lambda s}}_{z_{\text{hom}}},$$

where C is an arbitrary constant.

Below's verification in Maple that the function z_{prt} above is indeed a particular solution of the ODE

$$z'(s) + \lambda z(s) = g(s)$$

(execute the command `restart`; if you were on something beforehand):

```
> z := s -> exp(-lambda*s) * int( exp(lambda*s)*g(s),s );
```

$$s \rightarrow e^{(-\lambda s)} \int e^{(\lambda s)} g(s)\, ds$$

```
> diff(z(s),s)+lambda*z(s)-g(s);
```

$$e^{-\lambda s} e^{\lambda s} g(s) - g(s)$$

```
> simplify( diff(z(s),s)+lambda*z(s)-g(s) );
```

$$0$$

Proposition 3.1.3. *The change of variables*

$$\begin{cases} s = ax + by, \\ t = bx - ay \end{cases}$$

transforms an inhomogeneous linear PDE

$$au_x + bu_y + cu = d(x, y), \tag{3.1.8}$$

where $a, b, c \in \mathbf{R}$, and $a^2 + b^2 \neq 0$, to the PDE

$$w_s + \lambda w = \tilde{d}(s, t), \tag{3.1.9}$$

where

$$\lambda = \frac{c}{a^2 + b^2}$$

and

$$\tilde{d}(s, t) = \frac{1}{\lambda} d\left(\frac{as + bt}{a^2 + b^2}, \frac{bs - at}{a^2 + b^2} \right);$$

in turn, the equation (3.1.9) is reducible to an ODE. The general solution of (3.1.9) is given by

$$w(s,t) = \underbrace{e^{-\Lambda s} \int e^{\Lambda s} \tilde{d}(s,t)ds}_{w_{prt}} + \underbrace{e^{-\Lambda s} f(t)}_{w_{hom}}$$

The general solution of the equation (3.1.8) is then obtained from the right-hand side of the last equation after replacing s with ax + by and t with bx − ay.

Proof. As we have seen above, given a C^1-function u, our change of variables $x, y \to s, t$ transforms (the function)

$$au_x + bu_y + cu$$

to (the function)

$$(a^2 + b^2)w_s + cw,$$

which we can express by writing that

$$au_x + bu_y + cu = (a^2 + b^2)w_s + cw$$

(recall that the rigorous way to treat this equality is to treat s, t as both variables *and* functions in x, y).

Further, recalling the back replacements

$$\begin{cases} x = \dfrac{as + bt}{a^2 + b^2}, \\ y = \dfrac{bs - at}{a^2 + b^2}, \end{cases}$$

we can write that

$$d(x,y) = d\left(\frac{as + bt}{a^2 + b^2}, \frac{bs - at}{a^2 + b^2} \right).$$

Summing up, we see that the change of variables, described in the conditions, transforms the original PDE

$$au_x + bu_y + cu = d(x,y)$$

to the PDE

$$(a^2 + b^2)w_s + cw = d\left(\frac{as + bt}{a^2 + b^2}, \frac{bs - at}{a^2 + b^2} \right).$$

After dividing both sides by $a^2 + b^2$, we get

$$w_s + \lambda w = \tilde{d}(s, t),$$

where

$$\tilde{d}(s, t) = \frac{1}{a^2 + b^2} d\left(\frac{as + bt}{a^2 + b^2}, \frac{bs - at}{a^2 + b^2} \right),$$

(as we have stated in the formulation of the proposition).

This equation is in general an *inhomogeneous* linear equation, and the corresponding homogeneous linear equation is

$$w_s + \lambda w = 0 \quad (\Longleftrightarrow \ w = w_{\text{hom}}(s, t) = f(t)e^{-\lambda s})$$

whose general solution is described in Proposition 3.1.2.

Now, recall the general principle (Proposition 1.2.2) on the structure of the solution set of an inhomogeneous linear partial differential equation:

$$\begin{bmatrix} \text{The General Solution} \\ \text{of an Inhomogeneous} \\ \text{Linear Equation} \end{bmatrix} =$$

$$\underbrace{\begin{bmatrix} \text{A Particular Solution} \\ \text{of This Inhomogeneous} \\ \text{Linear Equation} \end{bmatrix}}_{???} + \underbrace{\begin{bmatrix} \text{The General Solution} \\ \text{of the Corresponding} \\ \text{Homogeneous Equation} \end{bmatrix}}_{f(t)e^{-\lambda s}}$$

So to complete the proof, we have to find at least one solution of the inhomogeneous equation

$$w_s + \lambda w = \tilde{d}(s, t).$$

In view of Example 3.1.2, this may be the function

$$w_{\text{prt}}(s, t) = e^{-\lambda s} \int e^{\lambda s} \tilde{d}(s, t) ds.$$

Thus the general solution of the equation

$$w_s + \lambda w = \tilde{d}(s, t)$$

is

$$w(s, t) = w_{\text{prt}}(s, t) + w_{\text{hom}}(s, t)$$
$$= e^{-\lambda s} \int e^{\lambda s} \tilde{d}(s, t) ds + e^{-\lambda s} f(t)$$

which completes the proof. □

Example 3.1.3. Find the general solution of the linear PDE

$$-u_x + u_y - u = \exp(-x + y).$$

Solution. Just to repeat the main points, briefly: a first-order linear PDE

$$au_x + bu_y + cu = d(x, y)$$

with constant coefficients a, b, c, where $a^2 + b^2 \neq 0$, "goes" under the change of variables

$$\begin{cases} s = ax + by \\ t = bx - ay \end{cases} \quad \Longleftrightarrow \quad \begin{cases} x = \dfrac{as + bt}{a^2 + b^2} \\ t = \dfrac{bs - at}{a^2 + b^2}, \end{cases}$$

to the equation

$$(a^2 + b^2)w_s + cw = d(x, y),$$

where

$$u(x, y) = w(s, t)$$

and where we (initially) think of s, t as functions in x, y.

In our case,

$$a = -1, \quad b = 1, \quad c = -1, \quad d(x, y) = \exp(-x + y).$$

So our first step in the solution is the following.

Step 1. *Write down the necessary change of variables:*

$$\begin{cases} s = -x + y \\ t = x + y \end{cases} \quad \Longleftrightarrow \quad \begin{cases} x = \dfrac{-s + t}{2} \\ y = \dfrac{s + t}{2}. \end{cases} \tag{3.1.10}$$

Step 2. *Write down the equation to which the original is transformed under the change of variables* $x, y \to s, t$.

As we have said above, in the general case it is the equation

$$(a^2 + b^2)w_s + cw = d(x, y).$$

Next, speaking of the general case, we have then to replace x, y in $d(x, y)$ with their expressions via s and t.

In our particular case, we have the equation

$$2w_s - w = \exp(-x + y),$$

or, by (3.1.10), as $s = -x + y$,

$$2w_s - w = \exp(s) \iff w_s - \frac{1}{2}w = \frac{1}{2}\exp(s).$$

Step 3. *Find the general solution of the transformed equation.*
It is the linear inhomogeneous equation

$$w_s - \frac{1}{2}w = \frac{1}{2}\exp(s).$$

So we have

- to obtain the general solution $w_{\text{hom}}(s, t)$ of the correspond homogeneous equation

$$w_s - \frac{1}{2}w = 0 \iff \underbrace{[w(s,t)]}_{z(s)}_s - \frac{1}{2}\underbrace{w(s,t)}_{z(s)} = 0; \tag{3.1.11}$$

- to find a particular solution $w_{\text{prt}}(s, t)$ of the inhomogeneous equation.

As is demonstrated in (3.1.11), the homogeneous equation is reducible to the ODE

$$z'(s) - \frac{1}{2}z(s) = 0 \iff z(s) = C\exp\left(\frac{1}{2}s\right).$$

Therefore,

$$w_{\text{hom}}(s, t) = f(t)\exp\left(\frac{s}{2}\right),$$

where f is a continuously differentiable function, is the general solution of the homogeneous equation.

According to Proposition 3.1.3, a particular solution w_{prt} of an inhomogeneous equation

$$w_s + \lambda w = g(s, t)$$

can be given by the function

$$w_{\text{prt}}(s, t) = e^{-\lambda s}\int e^{\lambda s}g(s, t)ds,$$

which, in our case, when $g(s, t) = \frac{1}{2}\exp(s)$, produces the function

$$w_{\text{prt}}(s, t) = \exp\left(\frac{s}{2}\right)\int \exp\left(-\frac{s}{2}\right)\frac{1}{2}\exp(s)ds$$

$$= \exp(s).$$

Therefore,

$$w_s - \frac{1}{2}w = \frac{1}{2}\exp(s)$$
$$\updownarrow$$
$$w(s,t) = w_{\text{prt}}(s,t) + w_{\text{hom}}(s,t)$$
$$\updownarrow$$
$$w(s,t) = \exp(s) + f(t)\exp\left(\frac{s}{2}\right).$$

Step 4. *Obtain the general solution of the original equation by making back replacements $s, t \to x, y$.*

We therefore get that

$$u(x,y) =$$
$$// \quad w(s,t) = \exp(s) + f(t)\exp\left(\frac{s}{2}\right) \quad // =$$
$$\exp(-x+y) + f(x+y)\exp((-x+y)/2).$$

The converse is also true: it may be either verified directly (say, with Maple), or we can use Proposition 3.1.3. □

Exercises

(As suggested earlier, use Maple, or W|a, or any other math software of your choice to verify your solutions.)

1. Find the general solution of the linear PDE

$$12u_x - 5u_y = 0,$$

and a particular solution of this PDE satisfying the initial condition

$$u(x,0) = x^2 - x^5$$

for all $x \in \mathbf{R}$.

2. Solve the linear PDE

$$-u_x + 3u_y + u = \exp(-x+2y).$$

4 First-order semilinear equations

It can be expected that our next subject will be the study of the general first-order linear equations

$$a(x,y)u_x + b(x,y)u_y + c(x,y)u = d(x,y).$$

Well, we do begin with analysis of the general solution of a first-order homogeneous linear PDE

$$a(x,y)u_x + b(x,y)u_y = 0 \tag{4.0.1}$$

on a domain Ω in $\mathbf{R}^2$. We shall be able to demonstrate that under certain natural conditions on the domain Ω and on the coefficient functions $a(x,y)$ and $b(x,y)$, there is a C^1-invertible change of variables $x, y \to s, t$, which reduces the equation to the PDE

$$w_s = 0.$$

This will imply that if we take any continuous function $c(x,y,z) \in C(\Omega \times \mathbf{R})$, then the PDE

$$a(x,y)u_x + b(x,y)u_y = c(x,y,u)$$

on Ω, which keeps the left-hand side of the linear equation (4.0.1) and, due to that, is called *semilinear*, is reducible to an ODE.

Before moving on, we will quote one of the most important results from the theory of ODEs.

4.1 Peano–Pickard–Lindelöf theorem

Theorem (Peano–Pickard–Lindelöf). *Let*

$$y' = f(x,y) \tag{4.1.1}$$

be an ODE. Suppose that the function $f(x,y)$ and its partial derivative $f_y(x,y)$ are continuous on a domain Ω in $\mathbf{R}^2$. Then given any point $\begin{pmatrix} x_0 \\ y_0 \end{pmatrix} \in \Omega$, there is a solution $y = \varphi(x)$ of the ODE, whose domain $\mathrm{dom}(\varphi)$ is an open interval about x_0, such that:

(i) $\varphi(x_0) = y_0$ (existence);

(ii) if ψ is another solution of the ODE such that $\mathrm{dom}(\psi)$ is an open interval about x_0 and $\psi(x_0) = y_0$, then

$$\varphi(x) = \psi(x)$$

for all $x \in \mathrm{dom}(\varphi) \cap \mathrm{dom}(\psi)$ (uniqueness).

https://doi.org/10.1515/9783110677256-004

The existence part is due to Giuseppe Peano (1858–1932) and the uniqueness part is due to Charles Émile Picard (1856–1941) and Ernst Leonard Lindelöf (1870–1946).

Note that if $\varphi : I \to \mathbf{R}$, where I is an open interval in $\mathbf{R}$, is a solution of the ODE (4.1.1), then, necessarily,

$$\begin{pmatrix} x \\ \varphi(x) \end{pmatrix} \in \Omega$$

for all $x \in I$, and so the graph of φ on I must lie entirely in Ω (for, indeed, given $x \in I$, the value of the function $f(x, y)$ in the right-hand side of (4.1.1) must exist at $\begin{pmatrix} x \\ \varphi(x) \end{pmatrix}$).

Geometrically, for every point $\begin{pmatrix} x_0 \\ y_0 \end{pmatrix} \in \Omega$, there is a curve

$$y = \varphi(x),$$

where φ is a solution of the ODE, which passes through this point, and if two such curves/solutions φ, ψ have a common point

$$\begin{pmatrix} x_0 \\ \varphi(x_0) \end{pmatrix} = \begin{pmatrix} x_0 \\ \psi(x_0) \end{pmatrix}$$

in Ω they must coincide on an (nonempty) open interval containing x_0.

Next, we apply the existence part of the Peano–Pickard–Lindelöf theorem to obtain a useful result which will be required in this chapter and in Chapter 8. First, however, we recall an important notion from the theory of ODEs.

Definition (First integrals of first-order explicit ODEs). Given a domain Ω in $\mathbf{R}^2$ and a function $f \in C(\Omega)$, a function $F \in C^1(\Omega)$ is called a *first integral* of the ODE

$$y' = f(x, y),$$

on Ω if for every solution $y = \varphi(x)$ the ODE whose domain is an open interval in $\mathbf{R}$ and whose graph is contained in Ω, F is constant on the graph of φ:

$$F(x, \varphi(x)) = \text{const}, \quad (x \in \text{dom}(\varphi)). \quad \triangle$$

We shall see below how a similar idea is used to define first integrals for *systems* of ODEs. We further refer the reader to Proposition 1.3.1 from Chapter 1 to see that the function F which is described in the formulation of the proposition, is a first integral of the corresponding ODE.

Some comments are in order for the concept a first integral of an ODE. As the reader will perhaps recollect, the notion of the *general solution* of an ODE is often either not defined, or remains rather vague. Things are at their best, when say, given a first-order ODE written in the general implicit form

$$F(x, y, y') = 0,$$

it is possible to obtain its *complete integral* that is, to find a function f satisfying

$$F(x, y, y') = 0 \iff y = f(x, C),$$

where C is a constant running in a suitable subset $\mathcal{C}$ of **R**. For example,

$$xy' - 3y = 0 \iff y = Cx^3,$$

where C runs over **R**. However, the complete integrals are more often than not hard to come by. For instance, a careful application of Proposition 1.3.1 to the separable ODE

$$y' = -\frac{2x}{\frac{y}{y+1}}, \tag{4.1.2}$$

and to a domain Ω in $\mathbf{R}^2$ which does not meet both the line $y = 0$ and the line $y = -1$ implies that for every solution $y = \varphi(x)$ of the ODE on an open interval I whose graph is contained in Ω we have that

$$\varphi(x) - \ln(|\varphi(x) + 1|) + x^2 = C$$

for a suitable constant C, everywhere on I (proof?). In other words, the function

$$F(x, y) = y - \ln(|y + 1|) + x^2$$

on Ω is a first integral of the ODE (4.1.2). Furthermore, one quickly sees that recovering of a complete integral of the ODE (4.1.2) is a pointless task, and we should consider the fact that we do have a *description* of all possible solutions of the ODE in terms of suitable first integrals $F \in C^1(\Omega)$ as a satisfactory one.

Most unambiguously, the *general solution* of a given ODE can be regarded as the set $\{\varphi\}$ of all functions $\varphi : I \to \mathbf{R}$, defined on open intervals in **R**, which satisfy the ODE everywhere on their domains. Conveniently, if need be, we can always consider the restriction $\psi = \varphi|_J$ of a particular solution $\varphi : I \to \mathbf{R}$ to any open subinterval J of I, thereby getting a solution of the ODE; or we can study the problem of existence of an extension of a particular solution $\varphi : I \to \mathbf{R}$ of the ODE to a solution of the ODE on a larger open interval K in **R**, and so on.

Here again, speaking of explicit first-order ODEs

$$y' = f(x, y),$$

first integrals provide a uniform and efficient way to distinguish suitable subfamilies of the general solutions.

Proposition 4.1.1. *Suppose that Ω is a domain in $\mathbf{R}^2$ and a real-valued function f on Ω and its partial derivative f_y are both continuous on Ω. Then every first integral $F \in C^1(\Omega)$ of the ODE*

$$y' = f(x, y) \tag{4.1.3}$$

is a solution of the PDE

$$U_x + f(x, y)U_y = 0$$

on Ω, and vice versa.

Proof. ($\Rightarrow$). Suppose that a smooth function $\varphi : I \to \mathbf{R}$, where I is an open interval in $\mathbf{R}$, whose graph is contained in Ω, is a solution of the ODE (4.1.3). Then, by the definition of a first integral,

$$F(x, \varphi(x)) = \text{const}$$

everywhere on I, and hence

$$[F(x, \varphi(x))]' = F_x(x, \varphi(x)) + F_y(x, \varphi(x)) \cdot \underbrace{\varphi'(x)}_{f(x, \varphi(x))} = 0,$$

or

$$F_x(x, \varphi(x)) + f(x, \varphi(x))F_y(x, \varphi(x)) = 0 \tag{4.1.4}$$

for all $x \in I$.

By the existence part of the Peano–Pickard–Lindelöf theorem, for every point $\begin{pmatrix} x_0 \\ y_0 \end{pmatrix} \in \Omega$, there is a solution $\psi : J \to \mathbf{R}$ of the ODE, satisfying Graph(ψ) $\subseteq \Omega$ such that

$$y_0 = \psi(x_0).$$

By applying (4.1.4) to the function ψ, we then get that

$$F_x(x_0, y_0) + f(x_0, y_0)F_y(x_0, y_0) = 0.$$

Since, however, $\begin{pmatrix} x_0 \\ y_0 \end{pmatrix}$ is an arbitrary point of Ω, we obtain that

$$F_x(x, y) + f(x, y)F_y(x, y) = 0$$

everywhere on Ω, as claimed.

($\Leftarrow$). Conversely, if $F \in C^1(\Omega)$ is a solution of the PDE

$$U_x + f(x,y)U_y = 0$$

on Ω, then given any solution $\varphi : I \to \mathbf{R}$ of the ODE (4.1.3), where I is an open interval in $\mathbf{R}$, whose graph is contained in Ω, we get that

$$F_x(x, \varphi(x)) + f(x, \varphi(x))F_y(x, \varphi(x)) = 0,$$

or

$$F_x(x, \varphi(x)) + F_y(x, \varphi(x))\varphi'(x) = 0,$$

for all $x \in I$, whence

$$[F(x, \varphi(x))]' = 0,$$

everywhere on I, which means that the function $F(x, \varphi(x))$ is constant on I. $\qquad\square$

4.2 Method of characteristics for semilinear equations

Proposition 4.2.1 (Method of characteristics). *Let Ω be a domain in $\mathbf{R}^2$ and let $a, b \in C(\Omega)$. Consider the homogeneous linear PDE*

$$a(x,y)u_x + b(x,y)u_y = 0 \qquad\qquad (4.2.1)$$

on Ω and suppose that
    (a) *$a(x,y)$ is nonzero everywhere on Ω and the function*

$$\frac{b(x,y)}{a(x,y)}$$

is continuously differentiable on Ω;
    (b) *the ODE*

$$y' = \frac{b(x,y)}{a(x,y)}, \qquad\qquad (4.2.2)$$

which is called the characteristic ODE *of the PDE (4.2.1), possesses a first integral $T \in C^1(\Omega)$ on Ω such that*
    (c) *the change of variables*

$$\begin{cases} s = x, \\ t = T(x,y) \end{cases} \quad \Longleftrightarrow \quad \begin{cases} x = s, \\ y = Y(s,t) \end{cases}$$

is C^1-invertible on Ω and

(d) *the image* $\Lambda = \Phi(\Omega)$ *of the domain* Ω *under the map*

$$\Phi\begin{pmatrix} x \\ y \end{pmatrix} = \begin{pmatrix} x \\ T(x,y) \end{pmatrix}, \quad \begin{pmatrix} x \\ y \end{pmatrix} \in \Omega$$

is an s-simple (horizontally convex) domain in the st-plane.
   Then:
   (i) *the change of variables*

$$\begin{cases} s = x, \\ t = T(x,y) \end{cases}$$

transforms the PDE

$$a(x,y)u_x + b(x,y)u_y = 0$$

to the PDE

$$a(s, Y(s,t))w_s = 0$$

for the function $w = w(s,t)$, *where* $u(x,y) = w(s,t)$, *which is equivalent, by* (a), *to the PDE*

$$w_s = 0;$$

   (ii) *in effect, the general solution of the PDE*

$$a(x,y)u_x + b(x,y)u_y = 0$$

on Ω *is given by*

$$u(x,y) = //\quad w(s,t) = f(t) \quad // = f(T(x,y)),$$

where f is a continuously differentiable function;
   (iii) *the family of curves*

$$T(x,y) = C, \quad (C \in \mathbf{R})$$

is called the family of characteristic curves *of the PDE* (4.2.1) *and, by* (ii), *any solution of this PDE is constant on any characteristic curve.*

Some comments are worth making on the conditions and statements of Proposition 4.2.1. We shall use the notation and the objects introduced in the formulation of the proposition.

Importantly, condition (b) in the proposition is true, provided that there is a function $T \in C^1(\Omega)$ such that

$$T_x(x,y) = b(x,y) \quad \text{and} \quad T_y(x,y) = -a(x,y) \tag{4.2.3}$$

everywhere on Ω. Indeed, according to Proposition 1.3.1 from Chapter 1, the function T is then a first integral of the ODE (4.2.2) on Ω.

In particular, as we have to remember, a function T satisfying (4.2.3) exists, whenever the characteristic ODE

$$y' = \frac{b(x,y)}{a(x,y)}$$

is a *separable* one, that is, an equation which can be (re)written in the form

$$y' = \frac{b(x)}{a(y)}.$$

Second, we comment on the shape of the graph of a particular solution of the homogeneous linear equation (4.2.1) on Ω. (Recall that we have made a similar comment after the formulation of Proposition 3.1.1 for the situation, when $a(x,y)$ and $b(x,y)$ are constant functions on $\mathbf{R}^2$; please consider rereading if necessary).

Understandably, the characteristic (plane) curves

$$T(x,y) = C,$$

where $C \in T(\Omega)$, of the equation (4.2.1) on Ω, which are also called *characteristics* for brevity's sake, fill the whole of Ω, or, in other words, Ω is a disjoint union of the characteristic curves, since given any point $P = \begin{pmatrix} x_0 \\ y_0 \end{pmatrix}$ of Ω, the characteristic curve

$$T(x,y) = T(x_0, y_0)$$

goes through P. Further, by part (ii) of the proposition, any particular solution

$$u = u(x,y) = f\big(T(x,y)\big)$$

of the equation is constant on every characteristic curve. Thus we can say that the graph Γ of u is made up of *copies* of characteristic curves, or that Γ is swept by a moving (space) curve of the form

$$\begin{cases} T(x,y) = C, \\ z = f(C), \end{cases} \tag{4.2.4}$$

where C runs over $T(\Omega)$, for indeed each curve of the form (4.2.4) is a translation of the characteristic curve

$$\begin{cases} T(x,y) = C, \\ z = 0 \end{cases}$$

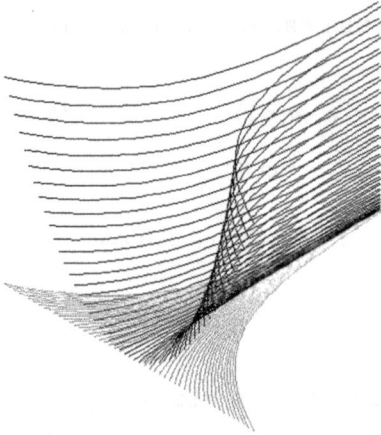

Figure 4.2.1: Characteristic curves and their translations.

in the plane Oxy along a direction parallel to the z-axis, and hence a copy of a characteristic curve of the equation (4.2.1). Figure 4.2.1 shows a number of the characteristic plane curves (shown in gray) associated with a certain PDE of the type under consideration, and their translations which lie on the graph of some solution of this PDE (shown in black).

Proof of Proposition 4.2.1. We have to perform a change of variables, and so have to choose whether to proceed with the direct, or the indirect method.

As before, as in the case of first-order linear equations with constant coefficients, we choose the *indirect* method.

Let $u = u(x, y)$ be a solution of the equation,

$$a(x, y)u_x + b(x, y)u_y = 0$$

on Ω. Our departure point is the identity

$$u(x, y) = w(s, t)$$

from which we deduce (by differentiating both parts first by x, and then by y) that

$$u_x = w_s s_x + w_t t_x,$$
$$u_y = w_s s_y + w_t t_y,$$

whence, from the definition of the new variables s and t ($s = x, t = T(x, y)$),

$$u_x = w_s + w_t T_x,$$
$$u_y = \quad\; w_t T_y.$$

We substitute the expressions for u_x and u_y in the left-hand side

$$a(x,y)u_x + b(x,y)u_y$$

of the original equation, thereby getting that

$$a(x,y)(w_s + w_t T_x) + b(x,y)w_t T_y$$
$$= a(x,y)w_s + \underbrace{(a(x,y)T_x + b(x,y)T_y)}_{\text{Zero?}} w_t.$$

The answer to the question in the last display is of course affirmative: by Proposition 4.1.1,

$$T_x + \frac{b(x,y)}{a(x,y)} T_y = 0$$

everywhere on Ω, whence, since the function $a(x,y)$ is never zero on Ω,

$$a(x,y)T_x + b(x,y)T_y = 0.$$

Thus our change of variables transforms the (original) equation

$$a(x,y)u_x + b(x,y)u_y = 0$$

to

$$a(x,y)w_s = 0.$$

As $a(x,y)$ is nonzero everywhere on Ω, the original equation on Ω is transformed to the PDE

$$w_s = 0$$

on $\Lambda = \Phi(\Omega)$. By (d) and by Proposition 1.4.1, the general solution of this equation on Λ is of the form

$$w(s,t) = f(t),$$

where f is a continuously differentiable function.

It follows that the general solution of the original equation

$$a(x,y)u_x + b(x,y)u_y = 0$$

on Ω is of the form

$$u(x,y) = /\!/ \quad w(s,t) = f(t) \quad /\!/ = f(T(x,y)).$$

All other statements follow easily. $\qquad\qquad\square$

As the reader may guess, verification of conditions (a)–(d) from Proposition 4.2.1 for a given linear homogenous equation

$$a(x,y)u_x + b(x,y)u_y = 0$$

on a certain domain Ω in $\mathbf{R}^2$ can be a rather difficult task. Verification of conditions (c) and (d) on C^1-invertibility of the required change of variables and on the shape of the transformed domain respectively can be particularly challenging. The following proposition, much like the Inverse Function Theorem for treating problems in its scope of applicability, can then be used to obtain a description of solutions at the *local* level. Recall that we have already met one local-level result (namely Corollary 1.4.2, which has a lot to do with the current topic), and, in fact, a particular result in the theory of PDEs is more likely to be a local-level one.

After studying the proof that follows, the reader is advised to think of possible applications and — no less importantly — misapplications of the proposition and of local-level results in general (and there will certainly be a chance to compare notes in the next chapter).

Proposition 4.2.2 (Method of characteristics: local version). *Let functions $a(x,y)$ and $b(x,y)$ on a domain Ω in $\mathbf{R}^2$ satisfy conditions* (a) *and* (b) *from Proposition 4.2.1. Suppose that a point $P = \begin{pmatrix} x_0 \\ y_0 \end{pmatrix}$ of Ω is such that*

$$T_y(x_0,y_0) \neq 0.$$

Then there is an open neighborhood $O_P \subseteq \Omega$ of the point P such that for every solution $u = u(x,y)$ of the PDE

$$a(x,y)u_x + b(x,y)u_y = 0$$

on Ω, there is a continuously differentiable function $f_P = f_P(t)$ on an open interval in $\mathbf{R}$ for which we have that

$$u(x,y) = f_P\bigl(T(x,y)\bigr)$$

for all points $\begin{pmatrix} x \\ y \end{pmatrix} \in O_P$.

Proof. As in the proof of Proposition 4.2.1, we consider the change of variables

$$\begin{cases} s = x, \\ t = T(x,y), \end{cases} \tag{4.2.5}$$

but, this time, first on an unspecified open neighborhood of the point P which is contained in Ω. The Jacobian matrix of the change is

$$\frac{\partial(s,t)}{\partial(x,y)} = \begin{pmatrix} s_x & s_y \\ t_x & t_y \end{pmatrix} = \begin{pmatrix} 1 & 0 \\ T_x(x,y) & T_y(x,y) \end{pmatrix},$$

and hence the Jacobian determinant is given by

$$\det \frac{\partial(s,t)}{\partial(x,y)} = \begin{vmatrix} 1 & 0 \\ T_x(x,y) & T_y(x,y) \end{vmatrix} = T_y(x,y).$$

Now as $T_y(x_0,y_0) \neq 0$, we have, by the Inverse Function Theorem, that there exists an open neighborhood $U_P \subseteq \Omega$ of the point P, which is a domain in $\mathbf{R}^2$, such that the change of variables (4.2.5) is C^1-invertible on U_P. Arguing as in the corresponding part of the proof of Proposition 4.2.1, we then get that the equation,

$$a(x,y)u_x + b(x,y)u_y = 0,$$

on U_P is transformed to the equation

$$w_s = 0$$

on the domain $\Phi(U_P)$ of the st-plane, where the map Φ on U_P is defined by

$$\Phi\begin{pmatrix} x \\ y \end{pmatrix} = \begin{pmatrix} x \\ T(x,y) \end{pmatrix}, \quad \begin{pmatrix} x \\ y \end{pmatrix} \in U_P.$$

Taking an open disk $D_Q \subseteq \Phi(U_P)$ centered at the point $Q = \Phi(P)$, which is convex, we see, by applying Proposition 1.4.1, that every solution of the PDE

$$w_s = 0$$

on D_Q is of the form

$$w(s,t) = f(t),$$

where f is a continuously differentiable function on an open interval in $\mathbf{R}$. This implies that any solution of the PDE

$$a(x,y)u_x + b(x,y)u_y = 0$$

on the open neighborhood $O_P = \Phi^{-1}(D_Q)$ of the point P is of the form

$$u(x,y) = /\!/ \quad w(s,t) = f(t) \quad /\!/ = f(T(x,y)),$$

and the result follows easily. □

It is hard to avoid a bad pun, but when we work with a given general first-order semilinear equation of the form

$$a(x,y)u_x + b(x,y)u_y = c(x,y,u),$$

Proposition 4.2.1 is only semi-helpful. Indeed, as the proof of our next result demonstrates, Proposition 4.2.1 can be used — in good cases — to reduce the equation to an ODE, but in general it cannot be used to uncover information as to the structure of the corresponding solution set.

Proposition 4.2.3. *Suppose that real-valued functions $a(x, y), b(x, y)$ on a domain $\Omega \subseteq \mathbf{R}^2$ satisfy conditions (a, b, c) from Proposition 4.2.1. Then the change of variables described in Proposition 4.2.1 transforms any first-order semilinear PDE of the form*

$$a(x, y)u_x + b(x, y)u_y = c(x, y, u), \tag{4.2.6}$$

where $c \in C(\Omega \times \mathbf{R})$, to the PDE

$$a(s, Y(s, t))w_s = c(s, Y(s, t), w),$$

and hence the equation (4.2.6) is reducible to an ODE.

The ordinary differential equation

$$y' = \frac{b(x, y)}{a(x, y)}$$

is called the *characteristic ODE* of the semilinear PDE (4.2.6).

Proof of Proposition 4.2.3. As we have demonstrated in the proof of Proposition 4.2.1, the change of variables

$$\begin{cases} s = x, \\ t = T(x, y) \end{cases}$$

transforms the function

$$a(x, y)u_x + b(x, y)u_y$$

to the function

$$a(x, y)w_s.$$

Therefore, under this change of variables, an equation of the form

$$a(x, y)u_x + b(x, y)u_y = c(x, y, u)$$

will be transformed to the equation

$$a(x, y)w_s = c(x, y, w).$$

We have to eliminate x, y from the last equation, and we can do this, since our change of variables is C^1-invertible on Ω:

$$\begin{cases} s = x, \\ t = T(x, y) \end{cases} \iff \begin{cases} x = s, \\ y = Y(s, t). \end{cases}$$

Thus we have that the original equation is transformed to the equation

$$a(s, Y(s, t))w_s = c(s, Y(s, t), w),$$

as claimed. □

Example 4.2.1. Find the general solution of the linear PDE

$$u_x + y u_y = 0$$

on $\Omega = \mathbf{R}^2$.

Solution. We shall apply Proposition 4.2.1, and so we have to verify all conditions from this proposition; here and below, we shall refer to them as conditions (a)–(d). The coefficient functions of the equation are the functions

$$a(x, y) = 1 \quad \text{and} \quad b(x, y) = y$$

on $\mathbf{R}^2$.

So condition (a) is true: indeed, the function

$$a(x, y) = 1 \neq 0$$

is nonzero everywhere on $\mathbf{R}^2$ and the function

$$\frac{b(x, y)}{a(x, y)} = y$$

is continuously differentiable on $\mathbf{R}^2$.

The characteristic ODE

$$y' = \frac{b(x, y)}{a(x, y)}$$

is in our case the equation

$$y' = \frac{y}{1} \iff y' = y, \tag{4.2.7}$$

for which we have

$$y' = y \iff y = Ce^x \iff \underbrace{ye^{-x}}_{T(x,y)} = C,$$

where $C \in \mathbf{R}$ is an arbitrary constant. The last equivalence shows that the function

$$T(x, y) = ye^{-x}$$

is a first integral of the characteristic ODE (4.2.7) on $\Omega = \mathbf{R}^2$. Let us then verify that conditions (c,d) are true for the function $T(x, y)$.

But first, as it is traditional (and instructive), we plot some of the characteristic curves

$$T(x, y) = C$$

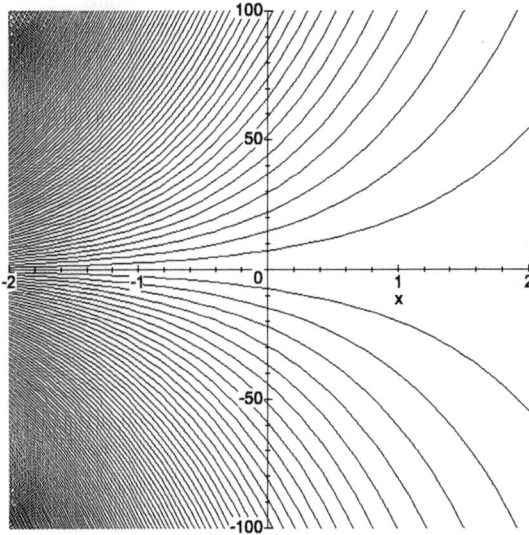

Figure 4.2.2: Example 4.2.1: characteristic curves $T(x,y) = ye^{-x} = C$ fill the whole of $\mathbf{R}^2$.

(see Figure 4.2.2). As it can be seen from Figure 4.2.2, the characteristic curves, as it was pointed out above, fill the whole of $\mathbf{R}^2$.

Further, the change of variables

$$\begin{cases} s = x, \\ t = ye^{-x} \end{cases} \quad \Longleftrightarrow \quad \begin{cases} x = s, \\ y = te^s \end{cases}$$

is C^1-invertible on $\mathbf{R}^2$, and so condition (c) also holds. Lastly, the image Λ of $\Omega = \mathbf{R}^2$ under the corresponding map

$$\Phi\begin{pmatrix} x \\ y \end{pmatrix} = \begin{pmatrix} x \\ ye^{-x} \end{pmatrix}, \quad \begin{pmatrix} x \\ y \end{pmatrix} \in \mathbf{R}^2$$

is equal to $\mathbf{R}^2$ itself (in other words, Λ is equal to the whole of the st-plane), and hence is s-simple. Thus condition (d) is true.

So our equation and the corresponding domain satisfy all the conditions of Proposition 4.2.1, and we conclude that the general solution of the equation on Ω is of the form

$$u(x,y) = f(T(x,y)) = f(ye^{-x}),$$

where f is a continuously differentiable function on $\mathbf{R}$. It is easy to verify that any function of this form is a solution of the original equation, which means that we did not make any mistakes when carrying out the solution... ☐

...but since mistakes are always possible, the final check must not, as a rule, be neglected.

Let us see what $W|\alpha$ has to "say': the query

```
D[ u(x,y),x ] + y*D[ u(x,y),y ]=0
```

brings us

$$u(x,y) = c_1(e^{-x}y),$$

as to be expected.

Example 4.2.2. Find the general solution of the linear equation

$$xu_x - yu_y + y^2u = y^2, \quad (x,y \neq 0).$$

Solution. This time, we shall apply Proposition 4.2.3. Since Proposition 4.2.3 is formulated for a *domain* in $\mathbf{R}^2$, that is, for an open and connected subset of $\mathbf{R}^2$, we derive from the conditions, which advise us to work with domains Ω in $\mathbf{R}^2$, all whose points $\binom{x}{y}$ satisfy

$$x,y \neq 0,$$

that we can work with one of the four maximal domains in $\mathbf{R}^2$ satisfying this requirement:

$$\Omega_1 = \left\{ \binom{x}{y} \in \mathbf{R}^2 : x > 0, y > 0 \right\},$$

$$\Omega_2 = \left\{ \binom{x}{y} \in \mathbf{R}^2 : x < 0, y > 0 \right\},$$

$$\Omega_3 = \left\{ \binom{x}{y} \in \mathbf{R}^2 : x < 0, y < 0 \right\},$$

$$\Omega_4 = \left\{ \binom{x}{y} \in \mathbf{R}^2 : x > 0, y < 0 \right\},$$

(in other words, with the interior of one of the standard four quadrants of $\mathbf{R}^2$), since the open set

$$\left\{ \binom{x}{y} \in \mathbf{R}^2 : x,y \neq 0 \right\}$$

in $\mathbf{R}^2$ is disconnected.

We shall work with $\Omega = \Omega_1$, the other cases being similar.

Further, the coefficients of our PDE, written in the standard form

$$a(x,y)u_x + b(x,y)u_y = c(x,y,u)$$

for first-order semilinear PDEs, are

$$a(x,y) = x, \quad b(x,y) = -y, \quad c(x,y,u) = y^2 - y^2 u.$$

Clearly, $a(x,y) = x$ is nowhere zero on Ω and the function $b(x,y)/a(x,y)$,

$$g(x,y) = \frac{b(x,y)}{a(x,y)} = -\frac{y}{x}$$

is continuously differentiable on Ω, since both the partial derivatives

$$g_x(x,y) = \left[-\frac{y}{x} \right]_x = \frac{y}{x^2}$$

and

$$g_y(x,y) = \left[-\frac{y}{x} \right]_y = -\frac{1}{x}$$

are continuous on Ω. Thus condition (a) is true.

Next, the characteristic ODE

$$y' = \frac{b(x,y)}{a(x,y)}$$

is the ODE

$$y' = -\frac{y}{x}. \tag{4.2.8}$$

Clearly, this ODE is separable, and hence we have that

$$\frac{dy}{y} = -\frac{dx}{x} \Rightarrow \ln|y| = -\ln|x| + C_0$$
$$\Rightarrow \ln|y| + \ln|x| = C_0$$
$$\Rightarrow \ln(|xy|) = C_0$$
$$\Rightarrow |xy| = \exp(C_0)$$
$$\Rightarrow xy = \pm\exp(C_0)$$
$$\Rightarrow xy = C,$$

where C is a constant (we hope that the reader mentally inserted the phrase "suppose each of the variables x, y varies in an open interval which does contain not 0" at the appropriate place in the sentence, and will continue this practice, whenever necessary). Conversely, if

$$y = \frac{C}{x}, \quad (x \neq 0)$$

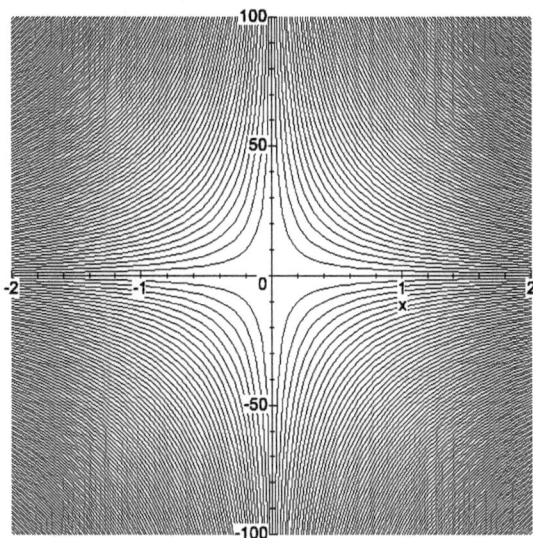

Figure 4.2.3: Characteristic curves filling $\mathbf{R}^2 - Ox - Oy$.

then

$$y' = -\frac{C}{x^2} \quad \text{and} \quad -\frac{y}{x} = -\frac{C}{x^2},$$

and hence $y(x)$ is a solution of (4.2.8). Thus

$$y' = -\frac{y}{x} \iff xy = C$$

on any open interval which does not contain 0, where C is an arbitrary nonzero constant. This implies that the function

$$T(x, y) = xy$$

is a first integral of the characteristic ODE on Ω, and that the equations of the characteristic curves are

$$T(x, y) = C \iff xy = C.$$

A number of the characteristics are shown in Figure 4.2.3. Again, the figure hints that through every point of Ω (of any domain Ω_k) passes a suitable characteristic curve.

Finally, the required change of variables,

$$\begin{cases} s = x, \\ t = xy \end{cases} \iff \begin{cases} x = s, \\ y = \dfrac{t}{s} \end{cases}$$

is C^1-invertible on Ω, and so condition (c) is also true.

We can then apply Proposition 4.2.3. Recall that it says that our change of variables transforms the original PDE

$$a(x,y)u_x + b(x,y)u_y = c(x,y,u)$$

to the PDE

$$a(s, Y(s,t))w_s = c(s, Y(s,t), w).$$

Simply speaking,

we have to replace x and y with their back replacements via s and t and u with w.

Thus

$$a(x,y) = //\quad a(s, Y(s,t))\quad // = x = s$$

and

$$c(x,y,u) = //\quad c(s, Y(s,t), w)\quad //$$
$$= y^2 - y^2 u$$
$$= \frac{t^2}{s^2} - \frac{t^2}{s^2} w$$

Therefore, the original equation is transformed to the equation

$$sw_s + \frac{t^2}{s^2} w = \frac{t^2}{s^2}$$

for the function $w = w(s,t)$ $//$ $= u(x,y)$ $//$.

What we have is an inhomogeneous linear equation. We have therefore

– to solve the corresponding homogeneous equation

$$sw_s + \frac{t^2}{s^2} w = 0;$$

– and to find a particular solution w_{prt} to the inhomogeneous equation.

The second task is easy: clearly,

$$w_{prt}(s,t) = 1$$

is a particular solution to the inhomogeneous equation.

Consider the homogeneous equation

$$sw_s + \frac{t^2}{s^2} w = 0.$$

As $s \neq 0$, it is equivalent to

$$w_s + \frac{t^2}{s^3} w = 0.$$ (4.2.9)

So let us solve an auxiliary ODE

$$z'(s) + g(s)z(s) = 0,$$

where g is a continuous function. We have

$$z'(s) + g(s)z(s) = 0 \Rightarrow \frac{dz}{z} = -g(s)ds$$

$$\Rightarrow \ln(|z|) = -\int g(s)\,ds + C_0$$

$$\Rightarrow |z| = \exp(C_0)\exp\left(-\int g(s)\,ds\right)$$

$$\Rightarrow |z| = \pm\exp(C_0)\exp\left(-\int g(s)\,ds\right)$$

$$\Rightarrow z = C\exp\left(-\int g(s)\,ds\right),$$

where, as it is easy to see, all the converse statements are also true, and hence the general solution of the ODE in question is

$$z(s) = C\exp\left(-\int g(s)ds\right).$$

This implies that the PDE (4.2.9) has the general solution

$$w_{\text{hom}}(s,t) = f(t)\exp\left(-\int \frac{t^2}{s^3}ds\right)$$

$$= f(t)\exp\left(\frac{t^2}{2s^2}\right)$$

Thus the general solution of the transformed equation is

$$w(s,t) = w_{\text{prt}}(s,t) + w_{\text{hom}}(s,t)$$

$$= 1 + f(t)\exp\left(\frac{t^2}{2s^2}\right).$$

The general solution of the original equation is therefore

$$u(x,y) = //\quad w(s,t) = 1 + f(t)\exp\left(\frac{t^2}{2s^2}\right)\quad //$$

$$= 1 + f(xy)\exp\left(\frac{y^2}{2}\right).$$

$\square$

As it is easy to verify, both Maple and $\mathbb{W}|\alpha$ agree with our solution. For instance, once entered at $\mathbb{W}|\alpha$, the query

```
x*D[ u(x,y),x ] -y*D[ u(x,y),y ] +y^2*u(x,y) =y^2
```

produces

$$u(x, y) = e^{y^2/2} c_1(xy) + 1.$$

Exercises

1. (i) Let $a, b \in \mathbf{R}$ be constants and let $a \neq 0$. Apply Proposition 4.2.1 to find the characteristic curves (characteristics) associated with the homogeneous linear PDE

$$au_x + bu_y = 0,$$

and its general solution on $\mathbf{R}^2$.

(ii) Let $a, b \in \mathbf{R}$ be a pair constants such that $b \neq 0$. How do you apply the method of characteristics in *this* situation to obtain the general solution of the PDE

$$au_x + bu_y = 0$$

on $\mathbf{R}^2$?

(iii) More generally, suppose that Ω is a domain in $\mathbf{R}^2$ and the coefficient functions $a(x, y), b(x, y)$ of the homogeneous linear PDE

$$a(x, y)u_x + b(x, y)u_y = 0$$

are C^1 on Ω and such that $b(x, y) \neq 0$ everywhere on Ω. Restate Proposition 4.2.1 for this case.

2. Find the characteristics, sketch the characteristics, perform the change of variables the method suggests, and then find the general solution of the linear PDE

$$xu_x + yu_y = x^n,$$

where $n > 0$ is a constant, on the domain

$$\Omega = \left\{ \begin{pmatrix} x \\ y \end{pmatrix} \in \mathbf{R}^2 : x > 0 \right\} = \mathbf{R}^+ \times \mathbf{R}$$

in $\mathbf{R}^2$. (*Hint*. To speed up the solution, look for a particular solution u_{prt} of the equation having the form $u_{\text{prt}}(x, y) = g(x)$.)

3. Solve the semilinear PDEs:

(i) $x^{10}u_x + xu_y = u$, $(x > 0)$;

(ii) $x^{10}u_x + xu_y = u^2$, $(x > 0)$;

(iii) $yu_x - xu_y = 0$, $(y > 0)$.

4. (i) Consider an explicit first-order ODE

$$y' = f(x, y)$$

on a domain $\Omega \subseteq \mathbf{R}^2$, where $f \in C^1(\Omega)$. Suppose that $F \in C^1(\Omega)$ is a first integral of this ODE such that

$$\nabla F = \mathrm{grad}(F) = \begin{pmatrix} F_x \\ F_y \end{pmatrix} \neq \begin{pmatrix} 0 \\ 0 \end{pmatrix}$$

everywhere on Ω. Prove that if $G \in C^1(\Omega)$ is another first integral of the ODE, then

$$G = \varphi_P(F),$$

where $\varphi_P = \varphi_P(s)$ is a continuously differentiable function, near every point $P \in \Omega$. (*Hint.* As, by Proposition 4.1.1, each of the functions F, G is a solution of the PDE

$$U_x + f(x, y)U_y = 0$$

on Ω, we have that

$$\begin{vmatrix} F_x & F_y \\ G_x & G_y \end{vmatrix} = 0$$

everywhere on Ω (why?). Apply then Proposition A.2.1).

(ii) Consider a homogeneous linear PDE

$$a(x, y)u_x + b(x, y)u_y = 0,$$

on a domain Ω in $\mathbf{R}^2$, where $a, b \in C(\Omega)$ and

$$\text{either } a(x, y) \neq 0, \quad \text{or} \quad b(x, y) \neq 0$$

at all points $\begin{pmatrix} x \\ y \end{pmatrix} \in \Omega$. Formulate and prove an analog of the statement in part (i) for solutions of the PDE on Ω.

5 First-order quasilinear equations: vector fields

In this and in the next two chapters, we shall study first-order *quasilinear* partial differential equations, equations that can be written in the form

$$a(x, y, u)u_x + b(x, y, u)u_y = c(x, y, u). \tag{5.0.1}$$

Clearly, any first-order linear, and any first-order semilinear PDE is quasilinear.

In this chapter, we shall introduce and study characteristic systems of ordinary differential equations associated with first-order quasilinear PDEs. The next chapter, Chapter 6, will be devoted to obtaining a description of solution sets of the first-order quasilinear PDEs, and in Chapter 7 we shall consider a method of construction of solutions of Cauchy problems involving first-order quasilinear PDEs (the last topic will also be studied in Chapter 6).

Let us discuss a natural condition — under which the equation (5.0.1) makes sense — on a domain Ω of the xy-plane (or, loosely speaking, on a domain Ω in $\mathbf{R}^2$) and on an open set Λ in $\mathbf{R}^3$, on which all functions $a(x, y, z)$, $b(x, y, z)$, and $c(x, y, z)$ participating in (5.0.1) are defined:

$$\Omega \subseteq \text{proj}_{x,y}(\Lambda), \tag{5.0.2}$$

where $\text{proj}_{x,y}$ is the orthogonal projection of $\mathbf{R}^3$ onto the xy-plane:

$$\text{proj}_{x,y}\begin{pmatrix} x \\ y \\ z \end{pmatrix} = \begin{pmatrix} x \\ y \end{pmatrix}, \qquad \begin{pmatrix} x \\ y \\ z \end{pmatrix} \in \mathbf{R}^3.$$

Recall that given a real-valued function u defined on a subset of $\mathbf{R}^2$, the *graph* of u is formally the set

$$\text{Graph}(u) = \left\{ \begin{pmatrix} x \\ y \\ u(x,y) \end{pmatrix} : \begin{pmatrix} x \\ y \end{pmatrix} \in \text{dom}(u) \right\}$$

of points of $\mathbf{R}^3$.

Why do we need the condition (5.0.2)? Well, for every solution $u \in C^1(\Omega)$ of the PDE (5.0.2) we must have that

$$\begin{pmatrix} x \\ y \\ u(x,y) \end{pmatrix} \in \Lambda \subseteq \text{dom}(a) \cap \text{dom}(b) \cap \text{dom}(c)$$

for all $\begin{pmatrix} x \\ y \end{pmatrix} \in \Omega$ (from now on, we will treat the subsets of the xy-plane in $\mathbf{R}^3$ as the subsets of $\mathbf{R}^2$, as suggested in (5.0.2)), and hence $\Omega \subseteq \text{proj}_{x,y}(\Lambda)$. To sum up, *the graph*

https://doi.org/10.1515/9783110677256-005

of u is contained in Λ. However, if some point $\begin{pmatrix} x_0 \\ y_0 \end{pmatrix} \in \Omega$ is not in the image $\text{proj}_{x,y}(\Lambda)$ of Λ, then there is no function $u \in C^1(\Omega)$, whose graph is contained in Λ, and accordingly, there is no function $u \in C^1(\Omega)$ which satisfies the equation (5.0.1).

5.1 Peano–Pickard–Lindelöf theorem for higher dimensions

Before studying the quasilinear equations themselves, we recall the general case of the Peano–Pickard–Lindelöf theorem.

What we have considered above was a particular case of this theorem for *real-valued* functions in one variable. However, speaking in more general terms, the theory of ODEs deals with *vector-valued* functions

$$t \mapsto \begin{pmatrix} \varphi_1(t) \\ \vdots \\ \varphi_n(t) \end{pmatrix},$$

where $\varphi_i(t)$ are continuously differentiable functions on a open subset of $\mathbf{R}$. It is rather straightforward to form a concept of an ODE for functions of this more general form. Geometrically, the functions of the above form are *smooth curves* in the space $\mathbf{R}^n$.

To make things less confusing, one usually speaks rather of *systems* of ODEs.

Theorem (Peano–Pickard–Lindelöf). *Let*

$$\begin{cases} y_1'(t) = f_1(t; y_1(t), y_2(t), \ldots, y_n(t)), \\ y_2'(t) = f_2(t; y_1(t), y_2(t), \ldots, y_n(t)), \\ \vdots \quad \vdots \quad \vdots \\ y_n'(t) = f_n(t; y_1(t), y_2(t), \ldots, y_n(t)) \end{cases} \tag{5.1.1}$$

be a system of n ordinary differential equations in n unknown functions $y_1, y_2, \ldots, y_n$.

Suppose that all the functions f_i *and all their partial derivatives* $[f_i]_{y_j}$ *are continuous on a domain* $\Omega \subseteq \mathbf{R}^{n+1}$. *Then given any point*

$$(t_0, y_1^0, y_2^0, \ldots, y_n^0)$$

of Ω,

(i) *there is a solution*

$$\Phi(t) = (\varphi_1(t), \varphi_2(t), \ldots, \varphi_n(t))$$

of the system, whose domain $\text{dom}(\Phi)$ *is an open interval about* t_0, *such that*

$$\varphi_1(t_0) = y_1^0, \ \varphi_2(t_0) = y_2^0, \ \ldots, \ \varphi_n(t_0) = y_n^0$$

(existence);

(ii) *if*

$$\Psi(t) = (\psi_1(t), \psi_2(t), \ldots, \psi_n(t))$$

is another solution such that $\mathrm{dom}(\Psi)$ *is an open interval about* t_0 *and*

$$\psi_1(t_0) = y_1^0, \; \psi_2(t_0) = y_2^0, \; \ldots, \; \psi_n(t_0) = y_n^0,$$

then

$$\Phi(t) = \Psi(t)$$

for every $t \in \mathrm{dom}(\Phi) \cap \mathrm{dom}(\Psi)$ (uniqueness).

Note, as we did earlier when quoting the theorem for the case when $n = 1$, that if $\Phi : I \to \mathbf{R}^n$, where I is an open interval in $\mathbf{R}$ and

$$\Phi(t) = (\varphi_1(t), \varphi_2(t), \ldots, \varphi_n(t)),$$

is a solution of the system (5.1.1), then the curve

$$(t, \Phi(t)) = (t, \varphi_1(t), \varphi_2(t), \ldots, \varphi_n(t))$$

on I must lie entirely in Ω.

Also, again as for the case when $n = 1$, geometrically, the existence part of the theorem means that given a point

$$P = (y_1^0, y_2^0, \ldots, y_n^0) \in \mathbf{R}^n,$$

such that $(t_0, P) \in \Omega$, there is a smooth curve

$$\Phi(t) = (\varphi_1(t), \varphi_2(t), \ldots, \varphi_n(t)),$$

a solution of the system (5.1.1), going through this point at $t = t_0$, and so

$$P = \Phi(t_0).$$

The uniqueness part means that if smooth curves $\Phi(t), \Psi(t)$, whose domains are open intervals in $\mathbf{R}$, are solutions the system (5.1.1), which take equal values at $t = t_0$,

$$\Phi(t_0) = \Psi(t_0),$$

then they coincide on the open interval $\mathrm{dom}(\Phi) \cap \mathrm{dom}(\Psi)$ about t_0.

We shall also need an important result on "continuous dependence on initial values," which is based on, and is an extension of the Peano–Pickard–Lindelöf theorem. Some preparatory work is in order.

We shall continue to follow the notation introduced in the formulation of the Peano–Pickard–Lindelöf theorem till the end of this section.

Let S denote the family of all solutions of the system (5.1.1) whose domains are open intervals in **R**. Given two solutions $\Phi_1, \Phi_2 \in S$, we say that Φ_2 is an *extension* of Φ_1, symbolically

$$\Phi_1 \preceq \Phi_2,$$

if $\operatorname{dom}(\Phi_1) \subseteq \operatorname{dom}(\Phi_2)$ and

$$\Phi_1(t) = \Phi_2(t)$$

for all $t \in \operatorname{dom}(\Phi_1)$. Observe for any $\Phi \in S$ we have that $\Phi \preceq \Phi$, that is, any element of S is formally an extension of itself (or, in other words, the relation $\preceq$ is reflexive). A solution Φ of the system (5.1.1) is called *nonextendable*, or a *maximal element* of S with respect to the relation $\preceq$, if

$$\Phi \preceq \Psi \Rightarrow \Psi = \Phi$$

for all $\Psi \in S$. The following lemma provides some basic properties of nonextendable solutions in S that are required in the theorem on continuous dependence below, and which are important in their own regard (particularly, for a better grasp of the concept of a nonextendable solution).

Lemma ([6, § 22A]). (i) *Given any point $\omega = (\tau, \xi_1, \ldots, \xi_n)$ of Ω, there is a nonextendable solution $\Phi_\omega \in S$ such that*

$$\Phi_\omega(\tau) = (\xi_1, \ldots, \xi_n);$$

(ii) *if Φ is a nonextendable solution in S and Ψ is any solution in S (resp., a nonextendable solution in S) such that*

$$\Phi(t) = \Psi(t)$$

for some point t in $\operatorname{dom}(\Phi) \cap \operatorname{dom}(\Psi)$, then $\Psi \preceq \Phi$ (resp., $\Psi = \Phi$).

Now take a point $\omega = (\tau, \xi_1, \ldots, \xi_n) \in \Omega$, and let $\Phi_\omega \in S$ be the (only) nonextendable solution such that

$$\Phi_\omega(\tau) = (\xi_1, \ldots, \xi_n).$$

Consider then a function $\boldsymbol{\varphi}$, whose domain is contained in the set $\mathbf{R} \times \Omega$, defined by the following condition:

$$\boldsymbol{\varphi}(t, \omega) = \Phi_\omega(t)$$

whenever $\omega \in \Omega$ and $t \in \operatorname{dom}(\Phi_\omega)$.

Theorem (on continuous dependence on initial values, [6, § 23, Theorem 14]). *The domain* $\mathrm{dom}(\boldsymbol{\varphi})$ *of the function* $\boldsymbol{\varphi}$ *defined above is an open subset of* $\mathbf{R} \times \Omega$*, and the function* $\boldsymbol{\varphi}$ *is continuous (at each point of its domain).*

Lev Pontryagin (1908–1988), the author of [6], urges the reader of his (on a similar occasion) to take good note of "nontriviality and importance" of the fact the domain of the function $\boldsymbol{\varphi}$ above is open.

Theorem (on differentiability with respect to initial values, [6, § 24, Theorem 17]). *Let* $(\tau_0, \xi_1^0, \ldots, \xi_n^0)$ *be a point of* Ω. *Consider the function*

$$\boldsymbol{\psi}(t, \xi_1, \ldots, \xi_n) = \boldsymbol{\varphi}(t, \tau_0, \xi_1, \ldots, \xi_n),$$

where the function $\boldsymbol{\varphi}$ *is defined above. Then the domain of the function* $\boldsymbol{\psi}$ *is an open subset of* $\mathbf{R}^{n+1}$*, and writing* $\boldsymbol{\psi}$ *as*

$$\boldsymbol{\psi}(t, \vec{\xi}) = (\psi_1(t, \vec{\xi}), \ldots, \psi_n(t, \vec{\xi})),$$

where $\vec{\xi} = (\xi_1, \ldots, \xi_n)$ *and where* ψ_i *are real-valued functions, we have that each of the component functions* ψ_i *is continuously differentiable by t and by each of the variables* ξ_j *and, moreover, the second-order partial derivatives*

$$[\psi_i]_{t\xi_j}, \quad (i, j = 1, 2, \ldots, n)$$

of the component functions are continuous everywhere on $\mathrm{dom}(\boldsymbol{\psi})$.

The last two theorems as presented in [6] are quite convenient to use and to apply. Unfortunately, there is no, best to the author's knowledge, English translation of [6] (the English translation [5] of an earlier Russian edition of Pontryagin's book does contain the theorems, but they are presented somewhat differently). A very useful, though again quite different, account on the theorems can be found in [9, Chapter III, § 13].

5.2 Vector fields and their characteristic curves

When dealing with a first-order quasilinear ODE

$$a(x, y, u)u_x + b(x, y, u)u_y = c(x, y, u), \tag{5.2.1}$$

where a, b, c are continuously differentiable functions, the system

$$\begin{cases} x'(t) = a(x(t), y(t), z(t)), \\ y'(t) = b(x(t), y(t), z(t)), \\ z'(t) = c(x(t), y(t), z(t)) \end{cases} \tag{5.2.2}$$

of ODEs is of utmost importance. The idea to associate this system of ODEs with the quasilinear equation (5.2.1) is due to Paul Charpit de Villecourt (?–1784), and it was further developed in the works by Joseph-Louis Lagrange (1736–1813), Gaspard Monge (1746–1818), Carl Gustav Jacob Jacobi (1804–1851), and other mathematicians.

In the theory of ODEs, systems of the form (5.2.2) are called *autonomous*, for we see that the functions a, b, c are functions in *three* variables, not in four variables, for example, like in

$$f_1(t, x(t), y(t), z(t)),$$

as one would normally expect.

In effect, the variable t is not that important (more on this below!), and that is why it is sometimes called a *dummy* variable.

Suppose that Λ is a domain in $\mathbf{R}^3$ on which the functions a, b, c are defined. Consider then the domain

$$\Omega = \mathbf{R} \times \Lambda \subseteq \mathbf{R}^4$$

lying in the space $\mathbf{R}^4$. Strictly formally, since the functions a, b, c are functions in three variables x, y, z, *not* in four variables, as is the case in the Peano–Pickard–Lindelöf theorem, we have that

the functions

$$a = a(x, y, z), \quad b = b(x, y, z), \quad c = c(x, y, z),$$

as well as all their partial derivatives, being considered as functions in four *variables* t, x, y, z *(in three "genuine" variables* x, y, z, *and in one dummy variable t) are — strictly formally — continuous on* $\Omega = \mathbf{R} \times \Lambda$.

Then, by the Peano–Pickard–Lindelöf theorem, given *any* $t_0 \in \mathbf{R}$ and any point $\begin{pmatrix} x_0 \\ y_0 \\ z_0 \end{pmatrix} \in \Lambda$, there is a solution $\begin{pmatrix} x(t) \\ y(t) \\ z(t) \end{pmatrix}$ of the system which passes through the point $P = \begin{pmatrix} x_0 \\ y_0 \\ z_0 \end{pmatrix}$ at $t = t_0$:

$$x(t_0) = x_0, \quad y(t_0) = y_0, \quad z(t_0) = z_0.$$

So as far as solutions of our system are concerned, the curves/solutions can pass through P at *any* possible t_0.

In particular, and it is very convenient, we can take $t_0 = 0$.

If we interpret the variable t as time (say, when doing physics), then P can be passed at any time. Accordingly, time is not that important in this regard, as discussed.

Definition (Vector fields, characteristic systems and characteristic curves of vector fields and quasilinear PDEs). Let Λ be an open set in $\mathbf{R}^3$ and let $a, b, c \in C^1(\Lambda)$ be functions that are continuously differentiable on Λ. Then:

(i) the function

$$\mathcal{F}_{a,b,c}\begin{pmatrix} x \\ y \\ z \end{pmatrix} = \begin{pmatrix} a(x,y,z) \\ b(x,y,z) \\ c(x,y,z) \end{pmatrix}$$

on Λ is called a *vector field* on Λ.

(ii) A smooth curve

$$y(t) = \begin{pmatrix} x(t) \\ y(t) \\ z(t) \end{pmatrix}, \quad (t \in I),$$

where I is an open interval in $\mathbf{R}$, which lies completely in Λ, that is,

$$y(t) \in \Lambda$$

for all $t \in I$, is called a *characteristic curve* of the vector field $\mathcal{F}_{a,b,c}$ if

$$\begin{cases} x'(t) = a(x(t), y(t), z(t)), \\ y'(t) = b(x(t), y(t), z(t)), \\ z'(t) = c(x(t), y(t), z(t)), \end{cases} \tag{5.2.3}$$

for all $t \in I$, or, for short,

$$y'(t) = \mathcal{F}_{a,b,c}(y(t))$$

for all $t \in I$. The system of ODEs (5.2.3) is called the *characteristic* system, or the system of the *Lagrange–Charpit equations* of the vector field $\mathcal{F}_{a,b,c}$.

We would like to stress that *any* open interval I can be the domain of a characteristic curve $y(t)$ provided that $y(I) \subseteq \Lambda$. Why it is so important? Well, while working with a particular characteristic curve $y(t)$ defined on an open interval I we may wish to switch to a smaller interval $J \subseteq I$ in order to have this or that property satisfied, thereby switching to a *new* curve $\eta = y|_J$, the restriction of y on J; observe that η is also a characteristic curve.

Incidently, as the reader undoubtedly knows, given a curve $y : I \to \mathbf{R}^3$ on an interval I, the set $y(I)$ is called the *range* of y. While it is OK to say that a curve $y : I \to \mathbf{R}^3$

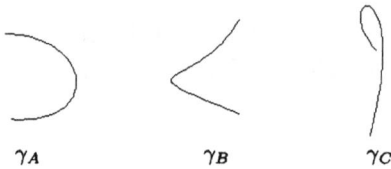

γ_A $\qquad\qquad$ γ_B $\qquad\qquad$ γ_C

Figure 5.2.1: Characteristic curves of the vector field $\mathcal{F}_{y,z,x}$.

is contained in this or that subset Λ of $\mathbf{R}^3$ (an extremely common, and thus forgivable, mistake), the correct way of expressing it is to say that the range of y is contained in Λ.

(iii) Let $a, b, c \in C^1(\Lambda)$ be continuously differentiable functions on Λ and let Ω be a domain in $\mathbf{R}^2$ such that the pair (Ω, Λ) satisfies the condition (5.0.2). Then the characteristic system and the characteristic curves of the vector field $\mathcal{F}_{a,b,c}$ on Λ are also called the *characteristic system* and *characteristic curves*, or simply *characteristics* of the PDE

$$a(x, y, u)u_x + b(x, y, u)u_y = c(x, y, u)$$

on Ω. $\triangle$

Characteristic curves of vector fields have in general quite nontrivial structure. Say, we require a characteristic curve γ_P of a vector field $\mathcal{F}_{a,b,c}$ go through a point $P = \begin{pmatrix} x_0 \\ y_0 \\ z_0 \end{pmatrix}$ at $t = 0$. Then the curve γ_P may change quite dramatically as P changes (see Figure 5.2.1 which shows the characteristic curves γ_A, γ_B, and γ_C of the vector field $\mathcal{F}_{a,b,c}$, where $a(x, y, z) = y$, $b(x, y, z) = z$, and $c(x, y, z) = x$ determined by the points

$$A = \begin{pmatrix} 1 \\ 2 \\ 3 \end{pmatrix}, B = \begin{pmatrix} 2 \\ 3 \\ 1 \end{pmatrix}, \text{ and } C = \begin{pmatrix} 3 \\ 1 \\ 2 \end{pmatrix}, \text{ respectively}).$$

Recall that if a space curve

$$y(t) = \begin{pmatrix} x(t) \\ y(t) \\ z(t) \end{pmatrix}, \quad (t \in I),$$

is differentiable at $t = t_0$, then the vector

$$y'(t_0) = \begin{pmatrix} x'(t_0) \\ y'(t_0) \\ z'(t_0) \end{pmatrix}$$

is called the *tangent vector* to y at $y(t_0)$.

The line with the canonical equations

$$\frac{x - x(t_0)}{x'(t_0)} = \frac{y - y(t_0)}{y'(t_0)} = \frac{z - z(t_0)}{z'(t_0)}$$

is called the *tangent line* to y at the point $y(t_0)$.

Physics: if parameter t is time, then the tangent vector $\gamma'(t)$ (resp., the tangent line) shows the *direction* of movement along γ as time t changes.

Geometrically then, the fact that a given space curve

$$\gamma(t) = \begin{pmatrix} x(t) \\ y(t) \\ z(t) \end{pmatrix}$$

is a solution of an autonomous system

$$\begin{cases} x'(t) = a(x(t), y(t), z(t)) \\ y'(t) = b(x(t), y(t), z(t)) \\ z'(t) = c(x(t), y(t), z(t)) \end{cases}$$

$$\Updownarrow$$

$$\gamma'(t) = \mathcal{F}_{a,b,c}(\gamma(t))$$

means that if $\gamma(t)$ "arrives" at a point $P = \begin{pmatrix} x_0 \\ y_0 \\ z_0 \end{pmatrix}$ at $t = t_0$, then its tangent vector $\gamma'(t_0)$

(which, to repeat, shows the direction of the "movement" along γ), must be equal to

the value of our vector field $\mathcal{F}_{a,b,c}$ at $P = \begin{pmatrix} x_0 \\ y_0 \\ z_0 \end{pmatrix}$:

$$\gamma'(t_0) = \mathcal{F}_{a,b,c}(P) \iff \gamma'(t_0) = \mathcal{F}_{a,b,c}(\gamma(t_0)).$$

Physics, again: if $\mathcal{F}_{a,b,c}$ is a *force* field, then its characteristic curves will show how a particle moves under the corresponding forces (when at a certain point P, the resulting force which acts on the particle is represented by the vector $\mathcal{F}_{a,b,c}(P)$).

Example 5.2.1. Let $x_0, y_0, z_0 \in \mathbf{R}$ be any real numbers. Find the characteristic curve $\gamma(t) = \begin{pmatrix} x(t) \\ y(t) \\ z(t) \end{pmatrix}$ of Burgers' equation

$$u_x + u u_y = 0$$

on $I = \mathbf{R}$ which passes through a point $P = \begin{pmatrix} x_0 \\ y_0 \\ z_0 \end{pmatrix}$ of $\mathbf{R}^3$ at $t = 0$:

$$\gamma(0) = P \iff \begin{pmatrix} x(0) \\ y(0) \\ z(0) \end{pmatrix} = \begin{pmatrix} x_0 \\ y_0 \\ z_0 \end{pmatrix}$$

$$\Longleftrightarrow \quad \begin{cases} x(0) = x_0, \\ y(0) = y_0, \\ z(0) = z_0. \end{cases}$$

Solution. As

$$a(x, y, u) = 1, \quad b(x, y, u) = u, \quad c(x, y, u) = 0,$$

the characteristic system

$$\begin{cases} x'(t) = a(x(t), y(t), z(t)), \\ y'(t) = b(x(t), y(t), z(t)), \\ z'(t) = c(x(t), y(t), z(t)), \end{cases}$$

is

$$\begin{cases} x'(t) = 1, \\ y'(t) = z(t), \\ z'(t) = 0. \end{cases} \tag{5.2.4}$$

The general solution of the first equation is

$$x(t) = t + C_1.$$

The general solution of the third equation is

$$z(t) = C_3.$$

Hence the general solution of the second equation is

$$y'(t) = C_3 \Rightarrow y(t) = C_3 t + C_2.$$

Thus the general solution of the characteristic system is

$$y(t) = \begin{cases} x(t) = t + C_1, \\ y(t) = C_3 t + C_2, \\ z(t) = C_3. \end{cases}$$

Now we wish to understand what are C_1, C_2, C_3 if

$$y(0) = \begin{pmatrix} x_0 \\ y_0 \\ z_0 \end{pmatrix}.$$

Clearly,

$$C_1 = x_0, \quad C_2 = y_0, \quad \text{and} \quad C_3 = z_0.$$

Thus the characteristic curve of the Burgers' equation which passes through P at $t = 0$ is

$$y(t) = \begin{cases} x(t) = t + x_0, \\ y(t) = z_0 t + y_0, \\ z(t) = z_0. \end{cases}$$

□

Let us see how systems of ODEs satisfying specific initial conditions can be solved in Maple.

First, we "translate" the system into the Maple language:

```
> sys:=diff(x(t),t)=1, diff(y(t),t)=z(t), diff(z(t),t)=0;
```

$$sys := \frac{d}{dt}\, \mathrm{x}(t) = 1, \quad \frac{d}{dt}\, \mathrm{y}(t) = \mathrm{z}(t), \quad \frac{d}{dt}\, \mathrm{z}(t) = 0$$

To obtain the *general solution* of the system, we use the same command

```
dsolve
```

which we used earlier for solving ODEs:

```
>  dsolve({sys},{x(t),y(t),z(t)});
```

$$\{\mathrm{x}(t) = \_C1 + t,\ \mathrm{y}(t) = \_C2 + t\,\_C3,\ \mathrm{z}(t) = \_C3,\}$$

What is important for us is a possibility of not just obtaining the general solution of the system, but also a possibility of finding the unique solution (on $\mathbf{R}$) satisfying the conditions

$$x(0) = x_0, \quad y(0) = y_0, \quad z(0) = z_0,$$

that is, the *initial conditions* that are imposed on our ODEs.

Let us first try the initial conditions

$$x(0) = 1, \quad y(0) = 2, \quad z(0) = 3.$$

```
> dsolve( {sys, x(0)=1, y(0)=2, z(0)=3}, {x(t),y(t),z(t)} );
```

$$\{\mathrm{x}(t) = 1 + t,\ \mathrm{y}(t) = 2 + 3\,t,\ \mathrm{z}(t) = 3\}$$

and, finally, the general case:

```
dsolve( {sys, x(0)=x0, y(0)=y0, z(0)=z0}, {x(t),y(t),z(t)} );
```

$$\{x(t) = x0 + t,\ y(t) = y0 + t\,z0,\ z(t) = z0\}$$

Now let us solve the system with $W|\alpha$: the standard query is as follows:

```
{x'[t] == 1, y'[t] == z[t], z'[t] == 0,
 x[0] == x0, y[0] == y0,    z[0] == z0}
```

but, actually, $W|\alpha$ "understands" a far more simple natural input:

```
x'(t)=1, y'(t)=z(t), z'(t)=0,
x(0)=x0, y(0)=y0, z(0)=z0
```

which produces

$$x(t) = t + x0$$
$$y(t) = tz0 + y0$$
$$z(t) = z0$$

Recall that if a function $u = u(x,y)$ is differentiable at a point $\begin{pmatrix} x_0 \\ y_0 \end{pmatrix}$ of $\mathbf{R}^2$, the plane

$$z = z_0 + u_x(x_0, y_0)(x - x_0) + u_y(x_0, y_0)(y - y_0), \tag{5.2.5}$$

where $z_0 = u(x_0, y_0)$ is called the *tangent plane* to the graph of u at $\begin{pmatrix} x_0 \\ y_0 \\ z_0 \end{pmatrix}$. According to a basic fact from analytic geometry, a plane

$$Ax + By + Cz = D,$$

in $\mathbf{R}^3$, where A, B, C are constants, is perpendicular to the normal vector $\begin{pmatrix} A \\ B \\ C \end{pmatrix}$, and so we see that the tangent plane (5.2.5) is perpendicular to the vector $\begin{pmatrix} u_x(x_0, y_0) \\ u_y(x_0, y_0) \\ -1 \end{pmatrix}$, the corresponding normal vector.

The following proposition provides some basic facts on vector fields $\mathcal{F}_{a,b,c}$ and on the geometry associated with the corresponding first-order quasilinear equations.

Proposition 5.2.1. *Let Λ be an open set in $\mathbf{R}^3$, let $a, b, c \in C^1(\Lambda)$, and let t_0 be a real number. Then*

(i) *given any point $P = \begin{pmatrix} x_0 \\ y_0 \\ z_0 \end{pmatrix} \in \Lambda$, there is a characteristic curve of the vector field*

$\mathcal{F}_{a,b,c}$ *on Λ,*

$$\gamma(t) = \begin{pmatrix} x(t) \\ y(t) \\ z(t) \end{pmatrix}, \quad (t \in I),$$

where I is an open interval about t_0, which passes through P at $t = t_0$:

$$\gamma(t_0) = P \iff x(t_0) = x_0, y(t_0) = y_0, z(t_0) = z_0.$$

The uniqueness part of the Peano–Pickard–Lindelöf theorem implies that if η_1, η_2 are characteristic curves of the vector field $\mathcal{F}_{a,b,c}$ on Λ, whose domains are open intervals about t_0 and both passing through P at $t = t_0$, then

$$\eta_1(t) = \eta_2(t)$$

for all $t \in \mathrm{dom}(\eta_1) \cap \mathrm{dom}(\eta_2)$.

(ii) *Let Ω be a domain in $\mathbf{R}^2$ such that*

$$\Omega \subseteq \mathrm{proj}_{x,y}(\Lambda),$$

let $u = u(x, y) \in C^1(\Omega)$ be a solution of the quasilinear PDE

$$a(x, y, u)u_x + b(x, y, u)u_y = c(x, y, u) \tag{5.2.6}$$

on Ω, and let a point $P = \begin{pmatrix} x_0 \\ y_0 \\ z_0 \end{pmatrix} \in \Lambda$ be on the graph of u. Then the normal vector

$\begin{pmatrix} u_x(x_0, y_0) \\ u_y(x_0, y_0) \\ -1 \end{pmatrix}$ *to the graph of u at P is orthogonal to the field vector $\mathcal{F}_{a,b,c}(P) = \begin{pmatrix} a(P) \\ b(P) \\ c(P) \end{pmatrix}$*

at P, which means that the vector $\mathcal{F}_{a,b,c}(P)$ at P is parallel to the tangent plane to the graph of u at P, and so is the tangent vector $\gamma'(t_0)$ of any characteristics curve $\gamma(t)$ of the vector field $\mathcal{F}_{a,b,c}$ passing through P at $t = t_0$.

To rephrase the second statement in part (ii) of the proposition: the characteristic curve $y(t)$ is said to *touch* (or to *kiss*, as some would say) the graph of u at P. We shall see below that this statement can be considerably strengthened (the reader may try to guess how far, though it is not easy).

Proof of Proposition 5.2.1. (i) While Λ itself may not be a domain in $\mathbf{R}^3$, we can switch to an open ball $O_P \subseteq \Lambda$ centered at P and then apply the Peano–Pickard–Lindelöf theorem to the domain $\mathbf{R} \times O_P$ in $\mathbf{R}^4$. The result then follows easily.

(ii) Let a function $u = u(x, y)$ be a solution of the PDE (5.2.6) on Ω, and let a point $P = \begin{pmatrix} x_0 \\ y_0 \\ z_0 \end{pmatrix}$ belong to the graph of u. Consequently, $\begin{pmatrix} x_0 \\ y_0 \end{pmatrix} \in \Omega$ and $z_0 = u(x_0, y_0)$.

Further, since the function u is a solution of the PDE (5.2.6) on Ω, then

$$a(x_0, y_0, u(x_0, y_0))u_x(x_0, y_0) + b(x_0, y_0, u(x_0, y_0))u_y(x_0, y_0)$$
$$= c(x_0, y_0, u(x_0, y_0)),$$

or

$$a(x_0, y_0, z_0)u_x(x_0, y_0) + b(x_0, y_0, z_0)u_y(x_0, y_0)$$
$$= c(x_0, y_0, z_0),$$

or

$$a(P)u_x(x_0, y_0) + b(P)u_y(x_0, y_0) = c(P),$$

whence

$$\begin{pmatrix} a(P) \\ b(P) \\ c(P) \end{pmatrix} \cdot \begin{pmatrix} u_x(x_0, y_0) \\ u_y(x_0, y_0) \\ -1 \end{pmatrix} = 0,$$

where $\cdot$ in the left-hand side represents the dot product on $\mathbf{R}^3$. Therefore, the tangent plane

$$z - z_0 = u_x(x_0, y_0)(x - x_0) + u_y(x_0, y_0)(y - y_0)$$

to the graph of u at P must be parallel to the vector $\mathcal{F}_{a,b,c}(P)$.

Now if $y(t)$ is any characteristic curve of the vector field $\mathcal{F}_{a,b,c}$ passing through P at $t = t_0$,

$$y(t_0) = P,$$

we have, to repeat, that

$$y'(t_0) = \mathcal{F}_{a,b,c}(y(t_0)) = \mathcal{F}_{a,b,c}(P),$$

which completes the proof. $\qquad\square$

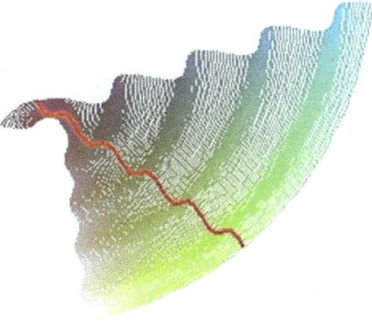

Figure 5.2.2: Characteristic curves passing through a given space curve.

Thus, given a vector fields $\mathcal{F}_{a,b,c}$ on an open set Λ in $\mathbf{R}^3$, we normally have a plethora of characteristic curves of this field with ranges in Λ (lying entirely in Λ).

In Chapter 7, we shall see that this enables us to make up surfaces out of these curves, thereby getting, under certain natural assumptions (of course), the graphs of particular solutions of the first-order quasilinear PDEs

$$a(x,y,u)u_x + b(x,y,u)u_y = c(x,y,u).$$

Let us briefly discuss the idea. Suppose functions a, b, c are continuously differentiable on Λ. To obtain a particular solution of the above PDE determined by the functions a, b, c, one usually starts with a smooth space curve Γ lying entirely in Λ and fixes some real number t_0 (say $t_0 = 0$). Let then I_P be an open interval about t_0 and let $\gamma_P(t) : I_P \to \Lambda$ be a characteristic curve of the corresponding vector field $\mathcal{F}_{a,b,c}$, whose range is contained in Λ, and which passes through a point $P \in \Lambda$ at $t = t_0$. We can then consider the (two-parametric) surface

$$\{\gamma_P : P \in \Gamma\},$$

which is the union (of the ranges) of the characteristic curves γ_P. An illustration to a setup like that is shown in Figure 5.2.2. To comment on the figure: the curve Γ is drawn in red, the surface in the figure,

$$\{\gamma_P : P \in \Gamma\},$$

is a union of the characteristic curves γ_P of the (quite trivial) vector field $\mathcal{F}_{a,b,c}$, where

$$a(x,y,z) = y, \quad b(x,y,z) = z, \quad c(x,y,z) = x$$

and where P runs over Γ.

5.3 First integrals of vector fields

Recall from Chapter 4, how useful is the assumption that the characteristic ODE

$$y' = \frac{b(x,y)}{a(x,y)}$$

of a first-order semilinear equation

$$a(x,y)u_x + b(x,y)u_y = c(x,y,u)$$

on a domain $\Omega \subseteq \mathbf{R}^2$ possesses a first integral on Ω.

As we shall see below, a similar "state of affairs" is also useful when dealing with the characteristic system of a given vector field $\mathcal{F}_{a,b,c}$ on an open set $\Lambda \subseteq \mathbf{R}^3$.

Note that given a continuously differentiable function $G \in C^1(\Lambda)$ and a constant $C \in \mathbf{R}$, the set

$$\left\{ \begin{pmatrix} x \\ y \\ z \end{pmatrix} \in \Lambda : G(x,y,z) = C \right\},$$

a level set of the function G, is "normally," if nonempty, a surface in $\mathbf{R}^3$, and pause momentarily to imagine a smooth curve lying in the intersection of two smooth surfaces in $\mathbf{R}^3$.

A statement of the form

$$\begin{cases} x' = a(x,y,z), \\ y' = b(x,y,z), \\ z' = c(x,y,z), \end{cases} \iff \begin{cases} F_1(x,y,z) = C_1, \\ F_2(x,y,z) = C_2, \end{cases} \tag{5.3.1}$$

where $F_1, F_2 \in C^1(\Lambda)$ and C_1, C_2 are arbitrary constants, is deciphered as follows: whenever

$$y(t) = \begin{pmatrix} x(t) \\ y(t) \\ z(t) \end{pmatrix}, \quad (t \in I),$$

is a characteristic curve defined on open interval I, then

$$F_1(x(t), y(t), z(t)) \equiv \text{const},$$
$$F_2(x(t), y(t), z(t)) \equiv \text{const}$$

everywhere on I, and, conversely, given any constants $C_1, C_2 \in \mathbf{R}$ and any point P of the set

$$\ell = \left\{ \begin{pmatrix} x \\ y \\ z \end{pmatrix} \in \Lambda : \begin{matrix} F_1(x,y,z) = C_1, \\ F_2(x,y,z) = C_2 \end{matrix} \right\},$$

"normally" a smooth curve, there is a characteristic curve $y(t)$ of the vector field $\mathcal{F}_{a,b,c}$ on Λ defined on an open interval I which goes through P and which lies entirely in ℓ, that is, $y(t) \in \ell$ for all $t \in I$ and $P = y(t_0)$ for some $t_0 \in I$.

Definition (First integrals of vector fields and characteristic systems of first-order quasilinear PDEs). Let functions a, b, c be continuously differentiable on an open set Λ in $\mathbf{R}^3$.

A function $G \in C^1(\Lambda)$ is called a *first integral* of the corresponding vector field $\mathcal{F}_{a,b,c}$ on Λ (resp., a *first integral* of the characteristic system in (5.3.1)) if G is constant on any characteristic curve of the field $\mathcal{F}_{a,b,c}$ (resp., on any solution of the characteristic system in (5.3.1)). △

Keeping in mind the role played by first integrals in our study of semilinear equations, we can (reasonably) expect that the situation when the general solution of a particular characteristic system is described by means of two suitable first integrals — as in (5.3.1) — should be considered as a favorable one.

Example 5.3.1. Find a description of the general solution of the characteristic system of Burgers' equation

$$u_x + uu_y = 0$$

of the form (5.3.1).

Solution. As we have seen above, the general solution of the characteristic system

$$\begin{cases} x'(t) = 1, \\ y'(t) = z(t), \\ z'(t) = 0, \end{cases} \quad \text{or} \quad \begin{cases} x' = 1, \\ y' = z, \\ z' = 0, \end{cases}$$

in question is

$$\begin{cases} x(t) = t + C_1, \\ y(t) = C_3 t + C_2, \\ z(t) = C_3, \end{cases}$$

where C_1, C_2, C_3 are arbitrary constants.

We have to find two suitable first integrals of the system (to repeat: functions that are constant on the characteristic curves): we look for smooth functions G such that

$$G(x(t), y(t), z(t)) \equiv \text{const},$$

or

$$G(t + C_1, C_3 t + C_2, C_3) \equiv \text{const}$$

for all $t \in \mathbf{R}$. One such a function is obvious:

$$F_1(x, y, z) = z,$$

for which we have

$$F_1(t + C_1, C_3 t + C_2, C_3) = C_3$$

for all $t \in \mathbf{R}$, as required. Thinking of other ways to produce a constant function playing with the functions $x(t), y(t), z(t)$ above, one quickly comes to another first integral:

$$F_2(x, y, z) = y - xz.$$

Indeed,

$$F_2(t + C_1, C_3 t + C_2, C_3) = C_3 t + C_2 - (t + C_1)C_3$$
$$= C_2 - C_1 C_3,$$

or

$$F_2(t + C_1, C_3 t + C_2, C_3) \equiv \text{const},$$

identically on $\mathbf{R}$, as claimed.

Thus, in the above notation,

$$\begin{cases} x' = 1, \\ y' = z, \\ z' = 0. \end{cases} \Rightarrow \begin{cases} z = D_1, \\ y - xz = D_2, \end{cases}$$

where D_1, D_2 are arbitrary constants.

Conversely, let $D_1, D_2 \in \mathbf{R}$ be arbitrary and let

$$\ell = \left\{ \begin{pmatrix} x \\ y \\ z \end{pmatrix} \in \mathbf{R}^3 : z = D_1, y - xz = D_2 \right\}.$$

Now, given any point P of ℓ is there a characteristic curve going through P and lying entirely in ℓ? Indeed, there is: for instance, the range of the characteristic curve

$$\gamma(t) = \begin{cases} x(t) = t, \\ y(t) = D_1 t + D_2, \\ z(t) = D_1, \end{cases}$$

where t runs over the whole of $\mathbf{R}$, is equal to ℓ.

Summing up,

$$\begin{cases} x' = 1, \\ y' = z, \\ z' = 0, \end{cases} \quad \Longleftrightarrow \quad \begin{cases} z = D_1, \\ y - xz = D_2, \end{cases}$$

where D_1, D_2 are arbitrary constants.

It should be noted that the use of the general solution of the characteristic system for the problem at hand could be easily avoided (but hopefully it was helpful, since, first, the notion of a first integral is rather nontrivial, and second, since it is a quite legitimate method for finding first integrals). Indeed, we have that

$$\begin{cases} x' = 1, \\ y' = z, \\ z' = 0. \end{cases} \quad \Rightarrow z' = 0 \Rightarrow z \equiv \text{const}$$

and that

$$\begin{cases} x' = 1, \\ y' = z, \\ z' = 0. \end{cases} \quad \Rightarrow (y - xz)' = y' - x'z - xz'$$

$$= z - z = 0$$

$$\Rightarrow y - xz \equiv \text{const},$$

which again demonstrates that the functions

$$F_1(x, y, z) = z \quad \text{and} \quad F_2(x, y, z) = y - xz$$

are first integrals. $\hfill\square$

Now let us consider a vector field $\mathcal{F}_{a,b,c}$ on an open set $\Lambda \subseteq \mathbf{R}^3$, where the functions a, b, c are continuously differentiable on Λ, and the field vector

$$\begin{pmatrix} a(x, y, z) \\ b(x, y, z) \\ c(x, y, z) \end{pmatrix} \neq \begin{pmatrix} 0 \\ 0 \\ 0 \end{pmatrix}$$

is never zero on Λ. We wish to investigate the first integrals of the vector field $\mathcal{F}_{a,b,c}$ on Λ.

Take any point $P \in \Lambda$ and let

$$\gamma(t) = \begin{pmatrix} x(t) \\ y(t) \\ z(t) \end{pmatrix}, \quad (t \in I),$$

where I is an open interval in $\mathbf{R}$, be a characteristic curve of our vector field which passes through P at $t_0 \in I$: $P = y(t_0)$. Suppose a function $F(x, y, z) \in C^1(\Lambda)$ is a first integral of the vector field $\mathcal{F}_{a,b,c}$. Then

$$F(y(t)) = F(x(t), y(t), z(t)) \equiv \text{const}$$

everywhere on I. Differentiating both sides of the last identity by t, we get

$$F_x(y(t))x'(t) + F_y(y(t))y'(t) + F_z(y(t))z'(t) = 0,$$

for all $t \in I$, or, since $y(t)$ is a characteristic curve of the vector field,

$$\underbrace{a(y(t))}_{x'(t)} F_x(y(t)) + \underbrace{b(y(t))}_{y'(t)} F_y(y(t)) + \underbrace{c(y(t))}_{z'(t)} F_z(y(t)) = 0,$$

for all $t \in I$. In particular,

$$a(y(t_0))F_x(y(t_0)) + b(y(t_0))F_y(y(t_0))$$
$$+ c(y(t_0))F_z(y(t_0)) = 0,$$

or

$$a(P)F_x(P) + b(P)F_y(P) + c(P)F_z(P) = 0, \tag{5.3.2}$$

which means, since P is an arbitrary point of Λ, that the function F satisfies the PDE

$$a(x, y, z)U_x + b(x, y, z)U_y + c(x, y, z)U_z = 0 \tag{5.3.3}$$

on Λ.

Conversely, if a function $F \in C^1(\Lambda)$ is a solution of the PDE (5.3.3) and a smooth curve $y(t)$, whose domain is an open interval I in $\mathbf{R}$, is a characteristic curve of the vector field $\mathcal{F}_{a,b,c}$, then

$$a(y(t))F_x(y(t)) + b(y(t))F_y(y(t))$$
$$+ c(y(t))F_z(y(t)) = 0$$

for all $t \in I$. It follows that

$$F_x(y(t))x'(t) + F_y(y(t))y'(t) + F_z(y(t))z'(t) = 0$$

for all $t \in I$, and then

$$[F(y(t))]' \equiv 0,$$

whence, finally,

$$F(y(t)) \equiv \text{const}$$

identically on I, which means that F is a first integral of the vector field $\mathcal{F}_{a,b,c}$.

Next, assume that functions $U, F, G \in C^1(\Lambda)$ are any first integrals of the vector field $\mathcal{F}_{a,b,c}$ and that P is any point of Λ. Then, by (5.3.2),

$$a(P)U_x(P) + b(P)U_y(P) + c(P)U_z(P) = 0,$$
$$a(P)F_x(P) + b(P)F_y(P) + c(P)F_z(P) = 0,$$
$$a(P)G_x(P) + b(P)G_y(P) + c(P)G_z(P) = 0.$$

Since the field vector $\begin{pmatrix} a(P) \\ b(P) \\ c(P) \end{pmatrix}$ is never zero on Λ, we get, by Cramer's rule, that

$$\begin{vmatrix} U_x(P) & U_y(P) & U_z(P) \\ F_x(P) & F_y(P) & F_z(P) \\ G_x(P) & G_y(P) & G_z(P) \end{vmatrix} = 0$$

or, for short,

$$\begin{vmatrix} U_x & U_y & U_z \\ F_x & F_y & F_z \\ G_x & G_y & G_z \end{vmatrix} = 0.$$

Thus

$$\det \frac{\partial(U, F, G)}{\partial(x, y, z)} = 0$$

identically on Λ.

Now *if* we had had

$$\operatorname{rank} \frac{\partial(U, F, G)}{\partial(x, y, z)} \equiv 2$$

everywhere on Λ, we would have had, by Proposition A.2.2, that locally one of the functions U, F, G would have been a "function of two other functions" everywhere on Λ.

Better still, *if* the rank condition above had been true due to linear independence of the gradients ∇F and ∇G of F and G, respectively, everywhere on Λ, or, other words, *if* the functions F, G had been *functionally independent* on Λ,

$$\operatorname{rank} \begin{pmatrix} F_x & F_y & F_z \\ G_x & G_y & G_z \end{pmatrix} \equiv 2, \tag{5.3.4}$$

everywhere on Λ, then U would have been locally a function of the functions F, G everywhere on Λ, that is, for every $P \in \Lambda$, there would have been an open neighborhood $O_P \subseteq \Lambda$ of P and a continuously differentiable function $\varphi_P(s, t)$ such that

$$U \equiv \varphi_P(F, G)$$

identically on O_P.

Now the question is: are there vector fields having two functionally independent first integrals? The answer is yes, and it is given by the only vector field we have investigated so far: by the vector field associated with Burgers' equation. Indeed, according to Example 5.3.1, the functions

$$F(x, y, z) = z,$$

$$G(x, y, z) = y - xz$$

are first integrals of the corresponding vector field. We have

$$\text{rank} \begin{pmatrix} F_x & F_y & F_z \\ G_x & G_y & G_z \end{pmatrix} = \text{rank} \begin{pmatrix} 0 & 0 & 1 \\ -z & 1 & -x \end{pmatrix} = 2,$$

everywhere on $\Lambda = \mathbf{R}^3$, as required to provide an example we were looking for.

So let us assume that the vector field $\mathcal{F}_{a,b,c}$ we are working with does have two functionally independent first integrals F, G on Λ. Then, to revoke (5.3.2), we have

$$a(P)F_x(P) + b(P)F_y(P) + c(P)F_z(P) = 0$$

and

$$a(P)G_x(P) + b(P)G_y(P) + c(P)G_z(P) = 0$$

for all points $P \in \Lambda$. We therefore conclude that both gradient vectors $(\nabla F)(P)$ and $(\nabla G)(P)$ are perpendicular to the *nonzero* vector $\begin{pmatrix} a(P) \\ b(P) \\ c(P) \end{pmatrix}$. But then, since the vectors $(\nabla F)(P)$ and $(\nabla G)(P)$ are linearly independent, we get that the vector $\begin{pmatrix} a(P) \\ b(P) \\ c(P) \end{pmatrix}$ is *parallel* to the cross product $(\nabla F)(P) \times (\nabla G)(P)$ of the gradient vectors, or for short,

$$\begin{pmatrix} a \\ b \\ c \end{pmatrix} \parallel (\nabla F \times \nabla G)$$

everywhere on Λ.

The following proposition summarizes the work done (the reader is urged to review the above arguments, to see that each statement in the proposition, bar one which is easy, had a rigorous proof).

Proposition 5.3.1. *Let Λ be an open set in $\mathbf{R}^3$ and let a, b, c be continuously differentiable functions on Λ such that*

$$(a, b, c) \neq (0, 0, 0)$$

everywhere on Λ. Then:

(i) *every first integral $U = U(x, y, z)$ of the vector field $\mathcal{F}_{a,b,c}$ on Λ satisfies the PDE*

$$a(x, y, z)U_x + b(x, y, z)U_y + c(x, y, z)U_z = 0$$

on Λ, and vice versa;

(ii) *if the vector field $\mathcal{F}_{a,b,c}$ possesses two functionally independent first integrals F, G on Λ, then a function $U \in C^1(\Lambda)$ is a first integral of the vector field $\mathcal{F}_{a,b,c}$ if and only if for every point P of Λ, there are an open neighborhood $O_P \subseteq \Lambda$ of P and a continuously differentiable function $\varphi_P(s, t)$ such that*

$$U \equiv \varphi_P(F, G)$$

identically on O_P;

(iii) *if the vector field $\mathcal{F}_{a,b,c}$ possesses two functionally independent first integrals F, G on Λ, then the vectors $\begin{pmatrix} a \\ b \\ c \end{pmatrix}$ and $(\nabla F \times \nabla G)$ are parallel/proportional everywhere on Λ.*

Remark 5.3.2. (a) Part (ii) of the proposition describes, as is customary to say, the *general (first) integral* of the vector field $\mathcal{F}_{a,b,c}$ on Λ. As we have seen on similar occasions above, the statement has to be taken as it is, at its face value.

It is certainly *not* claimed (unlike what some textbooks have to say) that any first integral U of the vector field is a function of the form

$$U = \varphi(F, G),$$

where φ is a continuously differentiable function, on the *whole* of Λ, that is, that

~~U is globally a function of F, G on Λ~~ (strike that!)

(though, any such a function *is* a first integral); what the proposition has to say about U is that

U is locally *a function of F, G on Λ.*

We remind the reader that we have already met an example of such a situation above. Indeed (see Exercise 1.9 in Chapter 1), the general solution of the PDE

$$u_y = 0$$

on a domain Π in $\mathbf{R}^2$ is *not* in general written in the form

$$u(x, y) = f(x),$$

where f is a continuously differentiable function in one variable, but, certainly, by Corollary 1.4.2, any solution of this PDE on Π is *locally* of the above form.

(b) So, assuming the hypothesis of (ii), and, being on the safe side, we can state therefore that

the general solution, or better put, the solution set, of the PDE

$$aU_x + bU_y + cU_z = 0$$

on Λ is the family of C^1-functions that are locally functions of F, G on Λ.

The importance of this result lies in the fact that if Ω is any domain in $\mathbf{R}^2$ such that

$$\Omega \subseteq \mathrm{proj}_{x,y}(\Lambda),$$

then the set of solutions in $C^1(\Omega)$ of the first-order quasilinear PDE associated with the vector field $\mathcal{F}_{a,b,c}$ on Λ consists of C^1-functions on Ω that are locally determined by suitable first integrals of the vector field $\mathcal{F}_{a,b,c}$ on Λ (Proposition 6.1.1 in the next chapter; we shall work on the proof of the proposition first thing in the next chapter). To be more precise, Proposition 6.1.1 states that given any solution $u \in C^1(\Omega)$ of the PDE in question, for every point $P \in \Omega$, there exist an open neighborhood $O_P \subseteq \Omega$ of P and a suitable first integral H_P of the vector field, which has no critical points on Λ, such that

$$H_P(x, y, u(x, y)) = 0$$

identically on O_P and, moreover, H_P can be chosen as simple as $H_P = F - \varphi_P(G)$, or $H_P = G - \varphi_P(F)$, where φ_P is a continuously differentiable function in one variable. We see that what part (ii) of Proposition 5.3.1 and Proposition 6.1.1 have in common is that they are both "severely and irredeemably" local statements.

(c) To continue our discussion in (a): on a more practical footing, we may use the fact that, by Proposition 5.3.1, any first integral of the vector field is an *extension* to the whole of Λ of a continuously differentiable function $\varphi_P(F, G)$ defined on an open neighborhood $O_P \subseteq \Lambda$ of a point P of Λ, where φ_P is a continuously differentiable function (needless to say, not every such a "local integral" need be extendable to the whole of Λ, and the problem of describing of "extendable local integrals" may be very hard, etc.).

(d) Another problem with the mentioned result in the proposition is that finding two functionally independent first integrals may be at times tricky, as the reader may judge from our (very simple) example above concerning Burgers' equation. We shall consider one more problem on determining first integrals associated with a first-order quasilinear PDE to enhance the point.

(e) Our discussion in (a) is not aimed to discourage the reader, but merely to advise to be careful. More often than not, the results describing solution sets of the PDEs are quite complicated, and one must be very thorough when analyzing both the hypotheses and the conclusions (which may at times turn out to be rather difficult). $\triangle$

Example 5.3.2. Consider the first-order quasilinear PDE

$$(y + u)u_x + yu_y = x - y, \quad (y \neq 0).$$

By regarding the coefficient functions a, b, c as elements of $C^1(\Xi)$, where

$$\Xi = \left\{ \begin{pmatrix} x \\ y \\ z \end{pmatrix} \in \mathbf{R}^3 : y \neq 0 \right\},$$

find two first integrals F, G of the corresponding vector field $\mathcal{F}_{a,b,c}$ on Ξ, and then find an open subset Λ of Ξ on which F, G are functionally independent.

Solution. The vector field in question is determined by the functions

$$a(x, y, z) = y + z, b(x, y, z) = y,$$
$$c(x, y, z) = x - y$$

which we should regard, according to the conditions, as members of $C^1(\Xi)$ (and, in effect, forget/ignore that they can be defined elsewhere). Clearly,

$$\begin{pmatrix} a(x, y, z) \\ b(x, y, z) \\ c(x, y, z) \end{pmatrix} \neq \begin{pmatrix} 0 \\ 0 \\ 0 \end{pmatrix}$$

everywhere on Ξ.

The characteristic system for the vector field $\mathcal{F}_{a,b,c}$ is therefore

$$\begin{cases} x'(t) = y(t) + z(t), \\ y'(t) = y(t), \\ z'(t) = x(t) - y(t), \end{cases}$$

or, for short,

$$\begin{cases} x' = y + z, \\ y' = y, \\ z' = x - y. \end{cases} \tag{5.3.5}$$

Suppose that functions

$$x = x(t), \quad y = y(t), \quad \text{and} \quad z = z(t),$$

which are defined on an open interval I in $\mathbf{R}$ and such that

$$y(t) \neq 0 \tag{5.3.6}$$

for all $t \in I$, give a solution of the characteristic system (5.3.5); the inequality in (5.3.6) makes sure that the characteristic curve

$$\gamma(t) = \begin{pmatrix} x(t) \\ y(t) \\ z(t) \end{pmatrix}, \quad (t \in I),$$

lies entirely in Ξ.

For convenience's sake, we will temporarily drop references to the variable t. We derive from the second equation

$$y' = y$$

in our system that y is an exponential function of the form Ce^t, where C is a nonzero, by (5.3.6), constant. Next, we add the third equation to the first one, thereby getting that

$$(x + z)' = x + z,$$

and, similarly, $x + z$ is an exponential function of the form De^t. Accordingly, the function

$$h_1 = \frac{x + z}{y}$$

must be constant. Alternatively, which may also be considered as justification of our reasoning above, differentiate h_1 by t:

$$h_1' = \left(\frac{x + z}{y} \right)' = \frac{(x + z)'y - (x + z)y'}{y^2}$$
$$= \frac{(x + z)y - (x + z)y}{y^2} = 0.$$

Before moving on: suppose that

$$f' = g \quad \text{and} \quad g' = f,$$

where f, g are some differentiable functions in one variable. Then

$$(f^2 - g^2)' = 2ff' - 2gg' = 2fg - 2gf = 0.$$

Let us obtain a pair of functions like f, g above, related to our functions x, y, z: we have that

$$z' = x - y$$

from the third equation of the system (5.3.5), and, in turn,

$$(x - y)' = x' - y' = y + z - y = z.$$

Now as

$$(x - y)' = z \quad \text{and} \quad z' = x - y,$$

the function

$$h_2 = (x - y)^2 - z^2$$

is a constant function.

To sum up: given any characteristic curve

$$y(t) = \begin{pmatrix} x(t) \\ y(t) \\ z(t) \end{pmatrix}, \quad (t \in I)$$

of the vector field $\mathcal{F}_{a,b,c}$, lying entirely in Ξ, the functions

$$h_1(t) = \frac{x(t) + z(t)}{y(t)}$$

and

$$h_2(t) = \big(x(t) - y(t)\big)^2 - z^2(t)$$

are constant on the interval I.

Therefore, by the definition, the functions

$$F(x, y, z) = \frac{x + z}{y}$$

and

$$G(x, y, z) = (x - y)^2 - z^2$$

are first integrals of our vector field on Ξ.

Let us find out on which open subset Λ of Ξ the functions F, G are functionally independent. We have

$$\begin{pmatrix} F_x & F_y & F_z \\ G_x & G_y & G_z \end{pmatrix} = \begin{pmatrix} \dfrac{1}{y} & -\dfrac{x+z}{y^2} & \dfrac{1}{y} \\ 2x - 2y & -2x + 2y & -2z \end{pmatrix}.$$

An advice to the reader: on a similar occasion, the resulting matrix of the partial derivatives can be quickly verified with the use of Maple. Say, for the current problem we can use the following Maple program:

```
with(linalg):
A := vector( [(x+z)/y, (x-y)^2-z^2, 0] ):
jacobian(A, [x,y,z]);
```

We must of course *remove* the zero row from the output, which will be

$$
\begin{pmatrix}
\dfrac{1}{y} & -\dfrac{x+z}{y^2} & \dfrac{1}{y} \\
2x-2y & -2x+2y & -2z \\
0 & 0 & 0
\end{pmatrix}.
$$

Now, everywhere on Λ,

$$
\operatorname{rank}
\begin{pmatrix}
\dfrac{1}{y} & -\dfrac{x+z}{y^2} & \dfrac{1}{y} \\
2x-2y & -2x+2y & -2z
\end{pmatrix}
$$

$$
= \operatorname{rank}
\begin{pmatrix}
1 & -\dfrac{x+z}{y} & 1 \\
x-y & -x+y & -z
\end{pmatrix},
$$

after multiplying the first row of the matrix on the left through by a nonzero number y and after dividing the second row of this matrix through by 2. Further, we subtract from the second row of the last matrix the first row multiplied by $(x - y)$,

$$
\operatorname{rank}
\begin{pmatrix}
1 & -\dfrac{x+z}{y} & 1 \\
x-y & -x+y & -z
\end{pmatrix}
$$

$$
= \operatorname{rank}
\begin{pmatrix}
1 & -\dfrac{x+z}{y} & 1 \\
0 & \dfrac{(y-x)(y-x-z)}{y} & y-x-z
\end{pmatrix}.
$$

Now we conclude that

$$
\operatorname{rank}
\begin{pmatrix}
F_x & F_y & F_z \\
G_x & G_y & G_z
\end{pmatrix} = 2
$$

at a point $\begin{pmatrix} x \\ y \\ z \end{pmatrix} \in \Xi$ if and only if

$$
y - x - z \neq 0.
$$

Accordingly, if

$$\Lambda = \left\{ \begin{pmatrix} x \\ y \\ z \end{pmatrix} : y \neq 0, y - x - z \neq 0 \right\},$$

then the functions F and G are functionally independent on Λ. In other words, the restriction of the vector field $\mathcal{F}_{a,b,c}$ on Λ has two functionally independent first integrals, namely $F|_\Lambda$ and $G|_\Lambda$.

A concluding remark: note that if we had considered the vector field determined by the coefficient functions of the PDE in question on $\mathbf{R}^3$, then the function $F(x, y, z)$ would have been out of the game, simply because it is not defined everywhere on $\mathbf{R}^3$. This makes us concentrate on an open set Ξ in $\mathbf{R}^3$ on which both F, G exist, and, following this, find an open subset Λ of Ξ on which F, G are functionally independent. Briefly put

$$\mathbf{R}^3 \to \Xi \to \Lambda,$$

that is, from $\mathbf{R}^3$ to Ξ, and then from Ξ to Λ. $\qquad\square$

Exercises

1. Suppose that a system of n linear ODEs in n unknown functions

$$\begin{cases} x_1'(t) = a_{11}x_1(t) + \cdots + a_{1n}x_n(t), \\ \vdots \quad \vdots \quad \vdots \quad \vdots \\ x_n'(t) = a_{11}x_1(t) + \cdots + a_{nn}x_n(t), \end{cases}$$

where $a_{ij} \in \mathbf{R}$, is such that the matrix

$$A = \begin{pmatrix} a_{11} & a_{12} & \cdots & a_{1n} \\ a_{21} & a_{22} & \cdots & \vdots \\ \vdots & \vdots & \ddots & \vdots \\ a_{n1} & a_{n2} & \cdots & a_{nn} \end{pmatrix}$$

of the system is *diagonalizable*, that is, if there are n linearly independent eigenvectors

$$\vec{v}_1, \vec{v}_2, \ldots, \vec{v}_n \in \mathbf{R}^n$$

of A, which correspond to eigenvalues

$$\lambda_1, \lambda_2, \ldots, \lambda_n \in \mathbf{R},$$

of A, respectively $(A\vec{v}_i = \lambda_i \vec{v}_i)$. Show then that the general solution of the system is given by

$$
\begin{pmatrix} x_1(t) \\ x_2(t) \\ \vdots \\ x_n(t) \end{pmatrix} = C_1 e^{\lambda_1 t} \vec{v}_1 + C_2 e^{\lambda_2 t} \vec{v}_2 + \cdots + C_n e^{\lambda_n t} \vec{v}_n,
$$

where $C_1, \ldots, C_n$ are arbitrary constants. (*Hint.* Start with the case when A itself is diagonal, and then use the fact that an $n \times n$ matrix A is diagonalizable if and only if a suitable conjugate $T^{-1}AT$ of A is diagonal.)

2. Consider the quasilinear PDE

$$
(y + u)u_x - 4yu_y = x - y.
$$

Use the previous exercise to obtain the general solution of the corresponding characteristic system.

Hint for Exercises 3 and 4: facing difficulties with determination of first integrals, try, as we did in Example 5.3.1, to obtain and to analyze the general solution of the corresponding characteristic system (which you can obtain with the use of $W|\alpha$ and/or Maple).

3. For each of the of the following first-order linear PDEs:
(a) consider the equation as a quasilinear one, and write down the characteristic system of ODEs;
(b) find two functionally independent first integrals of the corresponding vector field on the specified open set Λ in $\mathbf{R}^3$:

(i) $u_x = 0, \Lambda = \mathbf{R}^3$;

(ii) $xu_x + yu_y = u, \Lambda = \left\{ \begin{pmatrix} x \\ y \\ z \end{pmatrix} : x \neq 0 \right\}$;

(iii) $xu_x - yu_y = 2x^2, \Lambda = \left\{ \begin{pmatrix} x \\ y \\ z \end{pmatrix} : x \neq 0 \right\}$;

(iv) $yu_x - xu_y = 0, \Lambda = \left\{ \begin{pmatrix} x \\ y \\ z \end{pmatrix} : x \neq 0, \text{ or } y \neq 0 \right\}.$

4. For each of the following first-order quasilinear PDEs:
(a) write down the characteristic system of ODEs and find two first integrals of the corresponding vector field;

(b) find an open set Λ in $\mathbf{R}^3$ on which the first integrals are functionally independent (follow the concluding remark at the end of the solution of Example 5.3.2 above, which can be summarized by the diagram

$$\mathbf{R}^3 \to \Xi \to \Lambda,$$

that is, first identify an open set Ξ in $\mathbf{R}^3$ such that the restriction of the corresponding vector field $\mathcal{F}_{a,b,c}$ on Ξ does have a pair F, G of first integrals, and then work to find an open subset Λ of Ξ on which F, G are functionally independent):

    (i) $uu_x + yu_y = x$;

   (ii) $uu_x - yu_y = x$;

  (iii) $u_x + uu_y = -u^2$;

  (iv) $u_x + uu_y = u$;

   (v) $x(y - u)u_x + y(x + u)u_y = (x + y)u$.

To verify your solutions: form a $W|a$-style query describing a particular PDE, and enter it at the main page of the Wolfram Alpha website (https://wolframalpha.com).

Hopefully, $W|a$ will respond with a (local) description of an arbitrary solution of the PDE of the form

$$c_1(F(x, y, u(x, y)), G(x, y, u(x, y))) = 0,$$

where $F(x, y, z)$ and $G(x, y, z)$ are functionally independent first integrals of the corresponding vector field (recall that we have discussed above, in part (b) of Remark 5.3.2, why solutions of first-order quasilinear PDEs can be described in the above form).

6 First-order quasilinear equations: solution sets

The main of result of this chapter, Theorem 6.2.1, states that if a smooth vector field $\mathcal{F}_{a,b,c}(\Lambda)$ on an open set Λ in $\mathbf{R}^3$ possesses two functionally independent first integrals F, G, then, under some natural conditions, every solution of the corresponding quasilinear PDE

$$a(x, y, u)u_x + b(x, y, u)u_y = c(x, y, u)$$

on a domain $\Omega \subseteq \mathbf{R}^2$ satisfying $\Omega \subseteq \mathrm{proj}_{x,y}(\Lambda)$ is locally determined by the first integrals F, G everywhere on Ω.

Then we demonstrate how Theorem 6.2.1 can be applied to obtain local solutions of a Cauchy problem involving a first-order quasilinear PDE.

6.1 Solutions determined by first integrals

Proposition 6.1.1. *Let Λ be an open set in $\mathbf{R}^3$ and let Ω be a domain in $\mathbf{R}^2$ such that $\Omega \subseteq \mathrm{proj}_{x,y}(\Lambda)$. Suppose that functions a, b, c are continuously differentiable on Λ and that*

$$(a, b, c) \neq (0, 0, 0)$$

everywhere on Λ. Then:

(i) if $H = H(x, y, z)$ is a first integral of the vector field $\mathcal{F}_{a,b,c}$, whose gradient ∇H never vanishes on Λ, and if a function $u = u(x, y) \in C^1(\Omega)$, whose graph is contained in Λ, are such that

$$H(x, y, u(x, y)) \equiv 0$$

identically on Ω, then the function u is a solution of the quasilinear PDE

$$a(x, y, u)u_x + b(x, y, u)u_y = c(x, y, u) \tag{6.1.1}$$

on Ω;

(ii) suppose that the vector field $\mathcal{F}_{a,b,c}$ possesses two functionally independent first integrals F, G on Λ and that a function $u = u(x, y) \in C^1(\Omega)$ is a solution of the quasilinear equation (6.1.1). Then for every point $P \in \Omega$, there exist an open neighborhood $O_P \subseteq \Omega$ of P and a continuously differentiable function $\varphi_P = \varphi_P(s)$ such that either

$$F(x, y, u(x, y)) \equiv \varphi_P(G(x, y, u(x, y))),$$

or

$$G(x, y, u(x, y)) \equiv \varphi_P(F(x, y, u(x, y)))$$

https://doi.org/10.1515/9783110677256-006

identically on O_P. Accordingly,

$$H_1(x, y, u(x, y)) \equiv 0$$

identically on O_P for an appropriate first integral H_1 of the vector field $\mathcal{F}_{a_1, b_1, c_1}$ on Λ_P, whose gradient ∇H_1 never vanishes on Λ_P, where

$$\Lambda_P = \Lambda \cap (O_P \times \mathbf{R})$$

and where a_1, b_1, c_1 are the restrictions of the functions a, b, c on Λ_1, respectively.

Proof. (i) We differentiate the identity

$$H(x, y, u(x, y)) \equiv 0$$

on Ω by x and by y, thereby getting that

$$H_x + H_z u_x = 0 \iff \begin{vmatrix} u_x & -1 \\ H_x & H_z \end{vmatrix} = 0$$

and

$$H_y + H_z u_y = 0 \iff \begin{vmatrix} u_y & -1 \\ H_y & H_z \end{vmatrix} = 0.$$

Furthermore,

$$\begin{cases} u_y H_x = -u_x u_y H_z, \\ u_x H_y = -u_x u_y H_z \end{cases} \Rightarrow u_y H_x = u_x H_y \Rightarrow \begin{vmatrix} u_x & u_y \\ H_x & H_y \end{vmatrix} = 0.$$

Consequently,

$$\begin{pmatrix} u_x \\ u_y \\ -1 \end{pmatrix} \times \begin{pmatrix} H_x \\ H_y \\ H_z \end{pmatrix} = \begin{vmatrix} \vec{\mathbf{i}} & \vec{\mathbf{j}} & \vec{\mathbf{k}} \\ u_x & u_y & -1 \\ H_x & H_y & H_z \end{vmatrix} = \begin{pmatrix} 0 \\ 0 \\ 0 \end{pmatrix},$$

which means that the vectors participating in the cross product are parallel/proportional. On the other hand, by part (i) of Proposition 5.3.1, the gradient vector $\nabla H = \begin{pmatrix} H_x \\ H_y \\ H_z \end{pmatrix}$ is always orthogonal to the field vector $\begin{pmatrix} a \\ b \\ c \end{pmatrix}$ on Λ. In effect, since the gradient vector ∇H is nonzero, the vectors $\begin{pmatrix} u_x \\ u_y \\ -1 \end{pmatrix}$ and $\begin{pmatrix} a \\ b \\ c \end{pmatrix}$ are orthogonal, whence

$$au_x + bu_y + c(-1) = 0 \Rightarrow au_x + bu_y = c$$

at every point of the form $\begin{pmatrix} x \\ y \\ u(x,y) \end{pmatrix}$, where $\begin{pmatrix} x \\ y \end{pmatrix}$ runs over Ω, which means that

$$a(x,y,u)u_x + b(x,y,u)u_y = c(x,y,u)$$

everywhere on Ω, as claimed.

(ii) Let $u = u(x,y)$ be a solution of the PDE

$$a(x,y,u)u_x + b(x,y,u)u_y = c(x,y,u)$$

on Ω. Then at every point $P = \begin{pmatrix} x \\ y \\ u(x,y) \end{pmatrix}$, where $\begin{pmatrix} x \\ y \end{pmatrix} \in \Omega$, the field vector $\begin{pmatrix} a \\ b \\ c \end{pmatrix}$ is

orthogonal to the vector $\begin{pmatrix} u_x \\ u_y \\ -1 \end{pmatrix}$ and, by part (iii) of Proposition 5.3.1, to each of the

linearly independent vectors ∇F and ∇G. In effect, the triple product of the vectors

$\begin{pmatrix} u_x \\ u_y \\ -1 \end{pmatrix}$, ∇F and ∇G is equal to zero:

$$\begin{vmatrix} u_x & u_y & -1 \\ F_x & F_y & F_z \\ G_x & G_y & G_z \end{vmatrix} = 0.$$

Next, consider the continuously differentiable functions

$$f(x,y) = F(x,y,u(x,y))$$

and

$$g(x,y) = G(x,y,u(x,y))$$

on Ω. It easy to see that

$$\det \frac{\partial(f,g)}{\partial(x,y)} = \begin{vmatrix} f_x & g_x \\ f_y & g_y \end{vmatrix} = \begin{vmatrix} F_x + F_z u_x & G_x + G_z u_x \\ F_y + F_z u_y & G_y + G_z u_y \end{vmatrix}$$

$$= (F_x + F_z u_x)(G_y + G_z u_y)$$

$$\qquad - (F_y + F_z u_y)(G_x + G_z u_x)$$

$$= -u_x \begin{vmatrix} F_y & F_z \\ G_y & G_z \end{vmatrix} + u_y \begin{vmatrix} F_x & F_z \\ G_x & G_z \end{vmatrix} + \begin{vmatrix} F_x & F_y \\ G_x & G_y \end{vmatrix}$$

$$= -\begin{vmatrix} u_x & u_y & -1 \\ F_x & F_y & F_z \\ G_x & G_y & G_z \end{vmatrix} = 0,$$

or

$$\det \frac{\partial(f,g)}{\partial(x,y)} = 0$$

identically on Ω.

Now let us take any point $P = \begin{pmatrix} x_0 \\ y_0 \end{pmatrix}$ of Ω and let M denote the point $\begin{pmatrix} x_0 \\ y_0 \\ u(x_0,y_0) \end{pmatrix}$.

We claim that at least one of the partial derivatives

$$f_x, f_y, g_x, g_y$$

is nonzero at P. Indeed, otherwise we have that

$$0 = f_x = F_x + F_z u_x \Rightarrow F_x = -F_z u_x, \tag{6.1.2}$$
$$0 = f_y = F_y + F_z u_y \Rightarrow F_y = -F_z u_y$$

at M, which means that

$$(\nabla F)(M) = \begin{pmatrix} F_x \\ F_y \\ F_z \end{pmatrix} = \begin{pmatrix} -F_z u_x \\ -F_z u_y \\ F_z \end{pmatrix}$$

Similarly, if both partial derivatives g_x and g_y of g are equal to zero at P, then

$$(\nabla G)(M) = \begin{pmatrix} G_x \\ G_y \\ G_z \end{pmatrix} = \begin{pmatrix} -G_z u_x \\ -G_z u_y \\ G_z \end{pmatrix}.$$

Observe also that if F_z had been equal to zero at M, we would have, by (6.1.2), that

$$F_z = 0 \Rightarrow F_x = F_y = 0,$$

and so $(\nabla F)(M)$ would have been equal to the zero vector, which is absurd. But then there is a nontrivial linear combination of vectors $(\nabla F)(M)$ and $(\nabla G)(M)$ which is equal to the zero vector:

$$G_z(\nabla F)(M) - F_z(\nabla G)(M) = \begin{pmatrix} 0 \\ 0 \\ 0 \end{pmatrix},$$

again contradicting functional independence of F, G on Λ.

So let us assume that, say the partial derivative g_x is nonzero at P. Then due to continuity of g_x on Ω, g_x is nonzero on open neighborhood U_P of P. Now we are in a position to apply Proposition A.2.1. According to the proposition, there exist a suitable continuously differentiable function $\varphi = \varphi(s)$ and a suitable open disk $O_P \subseteq U_P$ centered at P such that

$$f(x, y) = \varphi(g(x, y)),$$

or

$$F(x, y, u(x, y)) = \varphi(G(x, y, u(x, y))) \tag{6.1.3}$$

everywhere on O_P.

The function

$$H_1(x, y, z) = F(x, y, z) - \varphi(G(x, y, z))$$

on $\Lambda_1 = \Lambda \cap (O_P \times \mathbf{R})$ is evidently a first integral of the vector field $\mathcal{F}_{a_1, b_1, c_1}$ on Λ_1, where a_1 (resp., b_1, c_1) is the restriction of a (resp., b, c) on Λ_1, since

$$\begin{aligned}
a_1 [H_1]_x + &b_1 [H_1]_y + c_1 [H_1]_z \\
&= a(F_x - \varphi'(G)G_x) + b(F_y - \varphi'(G)G_y) \\
&\quad + c(F_z - \varphi'(G)G_z) \\
&= aF_x + bF_y + cF_z - \varphi'(G)(aG_x + bG_y + cG_z) = 0
\end{aligned}$$

everywhere on Λ_1. By (6.1.3),

$$H_1(x, y, u(x, y)) = 0$$

for all $\begin{pmatrix} x \\ y \end{pmatrix} \in O_P$. Finally, let us show that the gradient ∇H_1 of H_1 is nonzero everywhere on Λ_1. Indeed, suppose otherwise. Then

$$\begin{aligned}
F_x - \varphi'(G)G_x &= 0, \\
F_y - \varphi'(G)G_y &= 0, \\
F_z - \varphi'(G)G_z &= 0,
\end{aligned}$$

at some point M of Λ_1, which implies that the vector

$$\nabla F - \varphi'(G)\nabla G$$

is equal to the zero vector at M, yet again contradicting functional independence of the functions F and G. $\qquad\square$

Remark 6.1.2. Let us discuss the results we have obtained in Proposition 6.1.1, exercising proper caution and care, as we did in Remark 5.3.2. We shall follow the notation of the proposition.

(a) It is stated in part (i) of the proposition that some solutions of the quasilinear PDE

$$a(x, y, u)u_x + b(x, y, u)u_y = c(x, y, u) \tag{6.1.4}$$

on Ω are determined by first integrals H of the vector field $\mathcal{F}_{a,b,c}$ that have no critical points in Λ: whenever

$$H(x, y, u(x, y)) \equiv 0$$

on Ω for a function $u \in C^1(\Omega)$, or, equivalently, whenever

H vanishes identically on the graph of u,

then u is a solution of the PDE (6.1.4) on Ω. Observe that part (i) of the proposition does not state that the converse is true, and so *a priori* there may be solutions of the PDE on Ω that are not determined by the first integrals having no critical points.

(b) Assume now that the vector field $\mathcal{F}_{a,b,c}$ has two functionally independent first integrals F, G. In this case, a prominent position among the first integrals we have considered in (a) is occupied by the first integrals H of the vector field $\mathcal{F}_{a,b,c}$ which are globally functions of F and G on Λ,

$$H = \varphi(F, G),$$

where the function $\varphi = \varphi(s, t)$ has no critical points on its domain, that is,

$$(\varphi_s, \varphi_t) \neq (0, 0)$$

everywhere on $\mathrm{dom}(\varphi)$. Indeed, we have

$$\nabla H = \begin{pmatrix} H_x \\ H_y \\ H_z \end{pmatrix} = \begin{pmatrix} \varphi_s F_x + \varphi_t G_x \\ \varphi_s F_y + \varphi_t G_y \\ \varphi_s F_z + \varphi_t G_z \end{pmatrix}$$

$$= \varphi_s \begin{pmatrix} F_x \\ F_y \\ F_z \end{pmatrix} + \varphi_t \begin{pmatrix} G_x \\ G_y \\ G_z \end{pmatrix},$$

or

$$\nabla H = \varphi_s \nabla F + \varphi_t \nabla G$$

everywhere on Λ. Therefore, due to linear independence of the gradients of F and G everywhere on Λ and our assumption on φ, the gradient ∇H of H never vanishes on Λ.

Thus any C^1-function u on Ω which is a solution of the functional equation

$$\varphi(F(x,y,u(x,y)), G(x,y,u(x,y))) = 0 \qquad (6.1.5)$$

on Ω is a solution of the PDE (6.1.4). Again, no guarantee, by our discussion in (a), that the converse is true (unlike what more than a few textbooks claim). Certainly, it is a good idea to refrain, to be on the safe side, from referring to (6.1.5) as the general solution of the PDE (6.1.4).

(c) Part (ii) of the proposition is a converse-of-sorts for part (i), but of course not the proper converse *per se*. It is claimed that if $u \in C^1(\Omega)$ is a solution of the PDE (6.1.4), then u is "locally determined" by "local" first integrals having the form

$$F - \varphi_P(G),$$

or the form

$$G - \varphi_P(F)$$

everywhere on Ω.

(d) Strictly formally, Proposition 6.1.1 deals with the solutions $u \in C^1(\Omega)$

whose graph is contained in Λ,

and that is where a lot of *misuses* of results similar to the proposition originate. Say, recall Example 5.3.2 in which we considered the PDE

$$(y + u)u_x + yu_y = x - y.$$

The corresponding vector field *on* $\mathbf{R}^3$ is determined by the functions

$$a(x,y,z) = y + z, \quad b(x,y,z) = y,$$
$$c(x,y,z) = x - y$$

which are continuously differentiable on the whole $\mathbf{R}^3$. However, to uncover a pair of first integrals, we moved to the open subset

$$\Xi = \{(x,y,z)^t : y \neq 0\}$$

of $\mathbf{R}^3$ and discovered two first integrals F, G for the restriction of the original vector field *on* Ξ, one of which simply did not exist on $\mathbf{R}^3$. To ensure functional independence of F, G we further moved to the open subset

$$\Lambda = \{(x,y,z)^t : y \neq 0, y - x - z \neq 0\}$$

of Ξ.

Thus Proposition 6.1.1 enables us to describe solutions of the PDE on domains Ω in $\mathbf{R}^2$ satisfying $\Omega \subseteq \mathrm{proj}_{x,y}(\Lambda)$, whose graph is contained in Λ. As far as any *other* solutions of the PDE on Ω are concerned, namely, those solutions whose graph is *not* contained in Λ, Proposition 6.1.1 provides no information. $\triangle$

Now, having discussed some finer points related to Proposition 6.1.1, we are ready to formulate a general statement on solution sets of the first-order quasilinear PDEs.

6.2 Main theorem on solution sets

Theorem 6.2.1. *Let Λ be an open set in $\mathbf{R}^3$ and let Ω be a domain in $\mathbf{R}^2$ such that $\Omega \subseteq \text{proj}_{x,y}(\Lambda)$. Suppose that functions a, b, c are continuously differentiable on Λ, that*

$$(a, b, c) \neq (0, 0, 0)$$

everywhere on Λ, and that the vector field $\mathcal{F}_{a,b,c}$ has two functionally independent first integrals F, G on Λ.

Then a function $u \in C^1(\Omega)$, whose graph is contained in Λ, is a solution of the PDE

$$a(x, y, u)u_x + b(x, y, u)u_y = c(x, y, u)$$

on Ω if and only if for every point $P \in \Omega$, there are an open neighborhood $O_P \subseteq \Omega$ of P and a continuously differentiable function $\varphi_P(s)$ such that u satisfies either the functional equation

$$F(x, y, u(x, y)) = \varphi_P(G(x, y, u(x, y))),$$

or the functional equation

$$G(x, y, u(x, y)) = \varphi_P(F(x, y, u(x, y)))$$

on O_P.

Proof. Argue like in the proof of part (i) of Proposition 6.1.1 to prove the sufficiency part, and use part (ii) of the proposition to prove the necessity part. □

We suggest the reader to compose a remark of theirs as to cautious use of Theorem 6.2.1. Then the reader is kindly requested to add the following two considerations to the remark:
- there is no really good answer to the question why solving functional equations is any easier than solving partial differential equations;
- the "severely local" character of the conclusion of Theorem 6.2.1, makes domains Ω on which to consider the corresponding PDE rather unimportant (unless otherwise required/instructed, more on that below): what is really important is an open set Λ on which we consider the coefficient functions a, b, c. When working with a particular set of functions, we may be forced to switch to a smaller open set in $\mathbf{R}^3$ to ensure that two functionally independent first integrals can be found, etc.

Example 6.2.1. Let Ω be a domain in $\mathbf{R}^2$. Describe the solutions of Burgers' equation,

$$u_x + uu_y = 0,$$

on Ω.

Solution. The corresponding vector field is

$$\mathcal{F}_{a,b,c}\begin{pmatrix} x \\ y \\ z \end{pmatrix} = \begin{pmatrix} 1 \\ z \\ 0 \end{pmatrix},$$

and $\mathcal{F}_{a,b,c}$ never vanishes on the whole of $\Lambda = \mathbf{R}^3$.

Next, as we know (Example 5.3.1), the functions

$$F(x,y,z) = z \quad \text{and} \quad G(x,y,z) = y - xz$$

are functionally independent first integrals of the vector field on $\mathbf{R}^3$.

Thus, by Theorem 6.2.1, a function $u \in C^1(\Omega)$ is a solution of Burgers' equation on Ω if and only if for every point $P \in \Omega$, there are an open neighborhood $O_P \subseteq \Omega$ and a continuously differentiable function φ_P such u satisfies either the functional equation

$$F(x,y,u(x,y)) = \varphi_P(G(x,y,u(x,y))),$$

or the functional equation

$$G(x,y,u(x,y)) = \varphi_P(F(x,y,u(x,y)))$$

on O_P. Accordingly, u satisfies either the functional equation

$$u(x,y) = \varphi_P(y - xu(x,y)),$$

or the functional equation

$$y - xu(x,y) = \varphi_P(u(x,y))$$

on O_P.

Understandably, we can choose a (one-for-all) continuously differentiable function φ and consider the solutions $u \in C^1(\Omega)$ of the functional equation

$$u(x,y) = \varphi(y - xu(x,y)),$$

or of the functional equation

$$y - xu(x,y) = \varphi(u(x,y))$$

on the whole of Ω (if any).

Say, let $\varphi(s) = s^2$ and let $\Omega = \mathbf{R}^+ \times \mathbf{R}^+$. Then there are four solutions of Burgers' equation on Ω determined by the corresponding first integrals $F - G^2$ and $G - F^2$: two C^1-solutions of the functional equation,

$$\underbrace{y - xu(x,y) = u^2(x,y)} = \varphi(u(x,y)),$$

on Ω:

$$u_{1,2}(x,y) = \frac{-x \pm \sqrt{x^2 + 4y}}{2},$$

and two C^1-solutions of the functional equation

$$\underbrace{u(x,y) = (y - xu(x,y))^2} = \varphi(y - xu(x,y)),$$

on Ω:

$$u_{3,4}(x,y) = \frac{1 + 2xy \pm \sqrt{1 + 4xy}}{2x^2}.$$

This shows that even such "global-level" C^1-functions φ generate quite a lot of solutions of Burgers' equation, and that the structure of the set of solutions of the equation is rich and complicated.

For the future reference, to verify the formulas for the functions $u_{1,2}$ and $u_{3,4}$ above, we can use the following Maple commands:

```
restart;

solve( y-x*z = z^2, z );

solve( z = (y-x*z)^2, z );
```

It is also a good idea to verify with the use of Maple that each of the functions u_i is indeed a solution of Burgers' equation on Ω. □

Let us see what the very same Maple has to "say" about Burgers' equation. Maple responds to the command

```
pdsolve( diff( u(x,y),x ) +u(x,y)*diff( u(x,y), y )=0, u(x,y));
```

with

$$u(x,y) = \mathrm{RootOf}(-y + \_Z * x + \_F1(\_Z))$$

which apparently should be understood as a claim that $u(x, y)$ is locally a solution of the functional equation

$$-y + Z(x, y)x + \_F1(Z(x, y)) = 0,$$

where $\_F1$ is a continuously differentiable function, which is in fact one of the functional equations

$$u(x, y) = \varphi_P(y - xu(x, y)),$$
$$y - xu(x, y) = \varphi_P(u(x, y)),$$

(namely, the second one), we have seen above. Truly, *if φ_P is locally C^1-invertible, then both equations become equivalent* on O_P if we reduce O_P accordingly. However, in general φ_P need not be locally C^1-invertible (say, $\varphi_P(s) = s^2$ as in the example above). Even simpler, the zero function $u(x, y) \equiv 0$, an evident solution of Burgers' equation, does not satisfy the second functional equation

$$y - xu(x, y) = \varphi_P(u(x, y))$$

for all C^1-functions φ_P (but satisfies the first one, for $\varphi_P \equiv 0$). So we see that Maple's solution should be taken with a grain of salt. On the positive side, we are presented with two functionally independent first integrals of the corresponding vector field.

$W|\alpha$ responds to the query

```
D[ u[x,y],x ] + u[x,y]*D[ u[x,y],y ]=0
```

with

$$u(x, y) = c_1\left(x - \frac{y}{u(x, y)}\right)$$

(as of December 22, 2019). This apparently means that the functions

$$F_1(x, y, z) = z \quad \text{and} \quad G_1(x, y, z) = x - \frac{y}{z}$$

are functionally independent first integrals associated with Burgers' equation. We have not met the second function before, though it is indeed, as it is easy to see, a first integral... but of the *restriction* of the corresponding vector field on the open set $\{z \neq 0\}$ in $\mathbf{R}^3$. This is of course unnecessarily restrictive (and the attentive user may wonder why all solutions of Burgers' equation should never take the zero value). Note that a couple of years ago, $W|\alpha$ responded to the same query with

$$c_1(u(x, y), y - xu(x, y)) = 0,$$

which was a safe way to respond. The above criticism notwithstanding, math software such as Maple, or $W|\alpha$ is of great help most of the time, but exercising a little caution on the road goes a long way.

Let us consider one more example of application of Theorem 6.2.1.

Example 6.2.2. Consider the first-order quasilinear PDE

$$(y + u)u_x + yu_y = x - y, \quad (y \neq 0).$$

As we know (see Example 5.3.2), the vector field determined by the restrictions of the coefficient functions on

$$\Lambda = \left\{ \begin{pmatrix} x \\ y \\ z \end{pmatrix} \in \mathbf{R}^3 : y \neq 0, y - x - z \neq 0 \right\},$$

has two first integrals F, G which are functionally independent. Use this fact to describe solutions $u \in C^1(\Omega)$, where Ω is a domain in $\mathbf{R}^2$ satisfying

$$\Omega \subseteq \mathrm{proj}_{x,y}(\Lambda),$$

of the above PDE on Ω whose graph is contained in Λ.

It is a good idea to reread part (d) of Remark 6.1.2, before moving on to the solution.

Solution. The first integrals in question are, as we have seen in the solution to Example 5.3.2, the functions

$$F(x, y, z) = \frac{x + z}{y} \quad \text{and} \quad G(x, y, z) = (x - y)^2 - z^2$$

on Λ. By applying Theorem 6.2.1, we see that a function $u \in C^1(\Omega)$, whose graph is contained in Λ, is a solution of the PDE in question if and only if for every point $P \in \Omega$, there are an open neighborhood $O_P \subseteq \Omega$ and a continuously differentiable function φ_P such that u satisfies either the functional equation

$$(x - y)^2 - u^2(x, y) = \varphi_P\left(\frac{x + u(x, y)}{y}\right),$$

or the functional equation

$$\frac{x + u(x, y)}{y} = \varphi_P((x - y)^2 - u^2(x, y))$$

on O_P. □

Try the query

```
( y+u[x,y] ) * D[ u[x,y],x ] +y*D[ u[x,y],y ] =x-y
```

to see what *W*]*a* has to say about the equation we have worked with, and following this, enter the command

```
pdsolve( (y+u(x,y)) * diff( u(x,y),x )
        +y*diff( u(x,y),y ) =x-y, u(x,y) );
```

to see Maple's reaction.

6.3 Cauchy problem for a PDE

The reader should remember various initial,

$$y(a) = A_0, \quad y'(a) = A_1, \quad y''(a) = A_2, \dots$$

or boundary conditions

$$y(a) = A_0, \quad y(b) = B_0, \quad y'(a) = A_1, \quad y'(b) = B_1, \dots$$

that are used to single out a particular solution of a given ODE on an interval $[a, b]$ in $\mathbf{R}$.

Likewise, we may think of using some additional constraints to (try to) single out a particular solution among the solutions of a given PDE.

Definition (Cauchy problems for first-order PDEs and their solutions). Consider a first-order partial differential equation

$$\mathcal{E}(u) = 0 \tag{6.3.1}$$

for an unknown function $u = u(x, y)$ in two variables, defined on an open set $\mathcal{O}$ in $\mathbf{R}^2$.

(i) A *Cauchy problem* for the PDE (6.3.1) can be written as

$$\begin{cases} \mathcal{E}(u) = 0, \\ u(f(s), g(s)) = h(s), \quad (s \in I), \end{cases} \tag{6.3.2}$$

where f, g, h are real-valued functions on an interval I in $\mathbf{R}$ (most commonly, f, g, h are required be smooth on I).

(ii) Let $\Omega \subseteq \mathcal{O}$ be a domain in $\mathbf{R}^2$ which is contained in $\mathcal{O}$ and such that

$$\begin{pmatrix} f(s) \\ g(s) \end{pmatrix} \in \Omega$$

for all $s \in I$. A *solution* of the Cauchy problem (6.3.2) *on* Ω is a function $u \in C^1(\Omega)$ which satisfies the PDE (6.3.1) on Ω and satisfies the initial conditions

$$u(f(s), g(s)) = h(s)$$

for all $s \in I$.

In effect, we require that the graph of any solution $u \in C^1(\Omega)$ of the Cauchy problem (6.3.2) on Ω contain the range of the *initial curve*

$$\Gamma(s) = \begin{pmatrix} f(s) \\ g(s) \\ h(s) \end{pmatrix}, \quad (s \in I),$$

or, equivalently, we require u take the *prescribed values* $h(s)$ on the range of the *base curve*

$$\widetilde{\Gamma}(s) = \begin{pmatrix} f(s) \\ g(s) \end{pmatrix}, \quad (s \in I).$$

The pair $(\widetilde{\Gamma}, h)$ is often referred to as *Cauchy data*. Observe that the base curve $\widetilde{\Gamma}$ can be characterized as the orthogonal projection of the initial curve Γ on the xy-plane (on $\mathbf{R}^2$).

Less restrictively (which may be helpful in practice),

(iii) the set of *solutions* of the Cauchy problem (6.3.2) is the union of the sets of solutions of the problem (6.3.2) on all suitable domains Ω in $\mathbf{R}^2$.

Accordingly,

(iv) the Cauchy problem (6.3.2) *has no solution*, if its solution set is empty. △

The reader may now guess that adequate formal interpretations of phrases like "Cauchy problem has a unique solution," or "Cauchy problem has infinitely many solutions," one frequently meets in the literature, should be rather nontrivial (spoiler: a proper formalization uses *equivalence relations*). We will address this later — after obtaining one theoretical result and solving our first Cauchy problem.

In the next section, we shall discuss a method of solution of Cauchy problems for first-order quasilinear PDEs which is based on Proposition 6.1.1 and uses the technique of first integrals (and in the next chapter we shall discuss another method, the method of characteristics).

6.4 First integrals: up to the task once again

Let us briefly discuss the idea of the method of solution of Cauchy problems for first-order quasilinear equations using first integrals of the corresponding vector fields.

Consider an open set Λ in $\mathbf{R}^3$, functions $a, b, c \in C^1(\Lambda)$, and a smooth curve $\Gamma(s)$ on an open interval I in $\mathbf{R}$ whose range is contained in Λ.

Assume that the vector field $\mathcal{F}_{a,b,c}$ possesses two functionally independent first integrals F, G on Λ.

As we know, solutions of the first-order quasilinear PDE on appropriate domains Ω in $\mathbf{R}^2$ are locally determined by the first integrals of the vector field $\mathcal{F}_{a,b,c}$: given any such a solution $u \in C^1(\Omega)$, a certain "local part" of the graph of u is contained in the level set

$$H(x, y, z) = 0$$

of a suitable first integral H.

Now given a Cauchy problem for the quasilinear PDE

$$a(x, y, u)u_x + b(x, y, u)u_y = c(x, y, u),$$

which involves the curve Γ as an initial curve, an immediate task is

to find a first integral H whose level set $H(x, y, z) = 0$ contains the range of Γ.

Well, one of the ways to achieve that is to try to find a function $\varphi \in C^1(\mathbf{R})$ in one variable such that, say

$$F \circ \Gamma(s) = \varphi\big(G \circ \Gamma(s) \big)$$

for all $s \in I$. If so, the level set $H(x, y, z) = 0$ of the first integral

$$H(x, y, z) = F(x, y, z) - \varphi\big(G(x, y, z)\big)$$

on Λ contains the range of Γ: $H(\Gamma(s)) = 0$ for all $s \in I$.

The next task, to be addressed to thereafter, is to force the graphs of solutions of the corresponding PDE, determined (or locally determined) by H, to contain the range of Γ as well.

Global solutions. Let us see how the latter task can be accomplished at the *global scale* (Proposition 6.4.1 below). Later we shall see how this global scale result — following a well-known strategy in multivariable calculus — can be transferred to a *local scale* result, which has in fact a wider range of applications.

Proposition 6.4.1. *Let Λ be a z-simple domain in $\mathbf{R}^3$, let functions $a, b, c \in C^1(\Lambda)$ be continuously differentiable on Λ and such that $(a, b, c) \neq (0, 0, 0)$ everywhere on Λ, let F, G be two functionally independent first integrals of the vector field $\mathcal{F}_{a,b,c}$ on Λ, and let Ω be a domain in $\mathbf{R}^3$ such that $\Omega \subseteq \mathrm{proj}_{x,y}(\Lambda)$. Suppose that*

$$\Gamma(s) = \begin{pmatrix} f(s) \\ g(s) \\ h(s) \end{pmatrix}, \quad (s \in I)$$

is a smooth curve on an open interval I in $\mathbf{R}$ whose range is contained in Λ and whose orthogonal projection $\tilde{\Gamma}$ on $\mathbf{R}^2$ is contained in Ω. Consider the Cauchy problem of finding a solution $u \in C^1(\Omega)$ of the PDE

$$a(x, y, u)u_x + b(x, y, u)u_y = c(x, y, u) \tag{6.4.1a}$$

satisfying the initial condition

$$u(f(s), g(s)) = h(s) \tag{6.4.1b}$$

everywhere on I.

Suppose further that

(a) *there is a continuously differentiable function $\varphi \in C^1(\mathbf{R})$ such that*

$$F(f(s), g(s), h(s)) = \varphi(G(f(s), g(s), h(s))),$$

or, for short,

$$F \circ \Gamma(s) = \varphi(G \circ \Gamma(s))$$

everywhere on I;

(b) *the corresponding first integral*

$$H(x, y, z) = F(x, y, z) - \varphi(G(x, y, z))$$

satisfies the condition

$$H_z \neq 0$$

everywhere on Λ;

(c) *the functional equation*

$$H(x, y, u(x, y)) = 0,$$

or, equivalently, the functional equation

$$F(x, y, u(x, y)) = \varphi(G(x, y, u(x, y)))$$

for an unknown C^1-function u on Ω whose graph is contained in Λ has a (unique) solution, a function $v = v(x, y) \in C^1(\Omega)$.

Then the function $v = v(x, y)$ is a solution of the Cauchy problem (6.4.1).

Proof. A crucial fact we shall base the proof on is as follows. Suppose that $\begin{pmatrix} x_0 \\ y_0 \end{pmatrix}$ is a point of Ω. Consider the function

$$\alpha(t) = H(x_0, y_0, t)$$

whose domain $\{t\}$ is determined by the condition $\begin{pmatrix} x_0 \\ y_0 \\ t \end{pmatrix} \in \Lambda$. We claim that the function $\alpha(t)$ is one-to-one. Indeed, suppose that

$$\alpha(t_1) = \alpha(t_2) \tag{6.4.2}$$

for some distinct points $t_1, t_2 \in \text{dom}(\alpha)$. For instance, assume that $t_1 < t_2$. As Λ is

z-simple, for every $t \in [t_1, t_2]$, the point $\begin{pmatrix} x_0 \\ y_0 \\ t \end{pmatrix}$ is in Λ, whence we get that α is defined

and continuously differentiable on $[t_1, t_2]$. Due to (6.4.2), we obtain, by Rolle's theorem, that there is an inner point $t_0 \in (t_1, t_2)$ at which the derivative of α is equal to zero. On the other hand,

$$0 = \alpha'(t_0) = H_z(x_0, y_0, t_0),$$

which contradicts the condition that H_z is nonzero everywhere on Λ.

Observe that since, by (c), the function $v = v(x, y)$ is a solution of the functional equation

$$H(x, y, u(x, y)) = 0$$

on Ω, it is a solution of the PDE

$$a(x, y, u)u_x + b(x, y, u)u_y = c(x, y, u)$$

on Ω (Proposition 6.1.1). Finally, we have, by (a) and (c), that

$$0 = H(f(s), g(s), \underbrace{h(s)}_{t_1}) = H(f(s), g(s), \underbrace{v(f(s), g(s)}_{t_2}))$$

for all $s \in I$. But, as we have proved above, this implies that

$$h(s) = v(f(s), g(s))$$

for all $s \in I$, and we are done. $\square$

As is now customary, we suggest the reader to think about, before resuming the reading, how Proposition 6.4.1 can be used and, no less importantly, *misused*.

Now having done as suggested, the reader may well agree that:
- most of misuses of the proposition happen whenever all the conditions (numerous, surely) have not been verified with the utmost care (especially it concerns the condition (c));
- most of uses of the proposition occur on a *local* scale, since the hypotheses of the proposition are quite strong ("the elephant is too big to ..."). The idea behind those uses, the above-mentioned well-known strategy in multidimensional calculus, is that any local scale arrangement can be as well considered as a global scale arrangement, both terms "local" and "global," being relative ones. We shall see below how this idea works, as promised.

In the following example, we shall see how Proposition 6.4.1 can be applied to solution of Cauchy problems related to Burgers' equation, application of the proposition in this case being quite possible, since the corresponding vector field on the whole of $\mathbf{R}^3$ is very well-behaved: it has two first integrals that are functionally independent on the whole of $\mathbf{R}^3$.

Example 6.4.1. Find a solution of the Cauchy problem

$$\begin{cases} u_x + uu_y = 0, \\ u(s, s^2) = s, \quad (s > 0), \end{cases} \tag{6.4.3}$$

on an appropriate domain Ω in $\mathbf{R}^2$, and examine the matter of its uniqueness.

Solution. Let us leave a domain Ω in $\mathbf{R}^2$ unspecified for a while — we assume only that Ω contains the range of the base curve

$$\tilde{\Gamma}(s) = \begin{pmatrix} s \\ s^2 \end{pmatrix}, \quad (s > 0),$$

(a piece of the parabola $y = x^2$ in the right open half-plane).

Consider, following the notation of Theorem 6.4.1, the initial curve

$$\Gamma(s) = \begin{pmatrix} f(s) \\ g(s) \\ h(s) \end{pmatrix} = \begin{pmatrix} s \\ s^2 \\ s \end{pmatrix}$$

on $I = (0, +\infty)$.

As we know, the vector field associated with Burgers' equation has two first integrals

$$F(x, y, z) = z \quad \text{and} \quad G(x, y, z) = y - xz,$$

which are functionally independent on the whole of $\mathbf{R}^3$ (Example 5.3.1). Certainly then, the corresponding restrictions of F, G are functionally independent on

$$\Lambda = \Omega \times \mathbf{R},$$

a z-simple domain in $\mathbf{R}^3$.

Next, we have that

$$F \circ \Gamma(s) = F(\Gamma(s)) = F(s, s^2, s) = s$$

and

$$G \circ \Gamma(s) = G(\Gamma(s)) = G(s, s^2, s) = s^2 - s \cdot s = 0$$

for all $s \in I$. We see that there is *no C^1-function ψ on **R**such that*

$$s = \underbrace{F \circ \Gamma(s)}_{} = \psi(\,G \circ \Gamma(s)\,) = \psi(0)$$

everywhere on I, but certainly there is a function $\varphi \in C^1(\mathbf{R})$ such that

$$0 = \underbrace{G \circ \Gamma(s)}_{} = \varphi(\,F \circ \Gamma(s)\,) = \varphi(s)$$

everywhere on I. For instance, we can take the zero function

$$\varphi(x) = 0$$

on **R**. Thus we need to consider the first integral

$$H(x, y, z) = G(x, y, z) - \varphi(F(x, y, z)) = G(x, y, z).$$

As

$$G_z = [y - xz]_z = -x,$$

let us assume that Ω *does not meet the line $x = 0$.* Accordingly, we shall have that

$$G_z \neq 0$$

everywhere on $\Lambda = \Omega \times \mathbf{R}$. Finally, the only solution of the functional equation

$$G(x, y, u(x, y)) = 0 \iff y - xu(x, y) = 0$$

on Ω is the function

$$v(x, y) = \frac{y}{x}, \quad \begin{pmatrix} x \\ y \end{pmatrix} \in \Omega.$$

Thus we conclude that, by Proposition 6.4.1, the function

$$v(x, y) = \frac{y}{x}$$

is a solution of the Cauchy problem at hand on Ω.

Verification is straightforward. Keeping in mind similar tasks in the future, we also suggest the reader to type and execute the following Maple program which can be used for verification of solutions of Cauchy problems for first-order quasilinear PDEs (understandably, the version of the program below is for the problem we are working on):

```
restart;

# Coefficient functions of the PDE

a := (x,y,z) -> 1 ;
b := (x,y,z) -> z ;
c := (x,y,z) -> 0 ;

# Components of the initial curve

f := s -> s   ;
g := s -> s^2 ;
h := s -> s   ;

# Potential solution

u := (x,y) -> y/x;

# Solution of the PDE?

simplify( a(x,y,u(x,y))*diff(u(x,y), x)
            + b(x,y,u(x,y))*diff(u(x,y), y)
            -c(x,y,u(x,y))  );

# Initial conditions?

simplify( u(f(s),g(s)) - h(s) );
```

If the program is typed and executed correctly, the reader will see that the last two lines of the output are

$$0$$

$$0$$

which will mean that the function in question is indeed a solution of the PDE (as signified by the first zero) and satisfies the initial condition (as signalled by the second zero).

Now let us discuss the matter of uniqueness.

Claim. (i) *Let $u \in C^1(\Sigma)$, where Σ is a domain in $\mathbf{R}^2$ which contains the range of the base curve $\widetilde{\Gamma}$, be a solution of the Cauchy problem (6.4.3). Then there is a subdomain Π of Σ, which contains the range $\mathrm{rng}(\widetilde{\Gamma})$ of $\widetilde{\Gamma}$, such that*

$$u(x, y) = \frac{y}{x}$$

for all points $\begin{pmatrix} x \\ y \end{pmatrix} \in \Pi.$

(ii) *Let* $u_1 \in C^1(\Omega_1)$ *and* $u_2 \in C^1(\Omega_2)$ *be solutions of the Cauchy problem* (6.4.3). *Then there is a domain* Π *in* $\mathbf{R}^2$ *such that*

$$\operatorname{rng}(\tilde{\Gamma}) \subseteq \Pi \subseteq \Omega_1 \cap \Omega_2$$

and

$$u_1(x, y) = u_2(x, y)$$

for all $\begin{pmatrix} x \\ y \end{pmatrix} \in \Pi.$

The reader undoubtedly noticed a certain similarity between the statement in part (ii) of the claim and the uniqueness statement in the Peano–Pickard–Lindelöf theorem. Based on the statement in part (ii) of the claim, we shall introduce below the notion of *equivalence* of solutions of a given Cauchy problem, as promised.

Proof of the Claim. (i) Let $s > 0$. Write P for the point $\tilde{\Gamma}(s) = \begin{pmatrix} s \\ s^2 \end{pmatrix}$ of Σ. By Theorem 6.2.1, there exists an open disk $U_s \subseteq \Sigma$ of radius $\eta_s > 0$ centered at P and a continuously differentiable function φ_P such that either

$$y - xu(x, y) = \varphi_P(u(x, y)), \tag{6.4.4a}$$

or

$$u(x, y) = \varphi_P(y - xu(x, y)) \tag{6.4.4b}$$

for all $\begin{pmatrix} x \\ y \end{pmatrix} \in U_s$. We also require that the disk U_s not meet the line $x = 0$, and we can certainly make the radius η_s smaller, if it not the case.

Suppose first that (6.4.4a) is true for all points in U_s. Then for all $x > 0$ satisfying $\begin{pmatrix} x \\ x^2 \end{pmatrix} \in U_s$, we have

$$0 = \underbrace{x^2 - xu(x, x^2) = \varphi_P(u(x, x^2))}_{} = \varphi_P(x),$$

and so

$$\varphi_P(x) = 0.$$

Since the curve $\tilde{\Gamma}$ is continuous at s, there exists a positive $\varepsilon_s > 0$ such that

$$|x - s| < \varepsilon_s \Rightarrow \left|\begin{pmatrix} x \\ x^2 \end{pmatrix} - \begin{pmatrix} s \\ s^2 \end{pmatrix}\right| < \eta_s \Rightarrow \varphi_P(x) = 0 \qquad (6.4.5)$$

for all $x > 0$. Now as

$$u(P) = u(s, s^2) = s > 0$$

and as u is continuous at $P = \begin{pmatrix} s \\ s^2 \end{pmatrix}$, we can find a positive $\delta_s < \eta_s$ for which we have

$$\left|\begin{pmatrix} x \\ y \end{pmatrix} - \begin{pmatrix} s \\ s^2 \end{pmatrix}\right| < \delta_s \Rightarrow u(x, y) > 0$$

and

$$\left|\begin{pmatrix} x \\ y \end{pmatrix} - \begin{pmatrix} s \\ s^2 \end{pmatrix}\right| < \delta_s \Rightarrow |u(x, y) - s| < \varepsilon_s$$

$$\Rightarrow \varphi_P(u(x, y)) = 0 \quad //\text{by (6.4.5)}//$$

$$\Rightarrow u(x, y) = \frac{y}{x}$$

for all $\begin{pmatrix} x \\ y \end{pmatrix} \in U_s$. Accordingly, letting B_s denote the open disk of radius δ_s centered at P, we conclude that

$$u(x, y) = \frac{y}{x}$$

on B_s.

On the other hand, (6.4.4b) cannot be true for all points of the disk U_s, since then we would have that

$$x = \underline{u(x, x^2) = \varphi_P(x^2 - xu(x, x^2))} = \varphi_P(0),$$

or

$$\varphi_P(0) = x$$

for infinitely many $x > 0$ satisfying $\begin{pmatrix} x \\ x^2 \end{pmatrix} \in U_s$ (we see that the argument essentially goes along the same lines as before, with the stress on the word "essentially").

We therefore see that

$$u(x, y) = \frac{y}{x}$$

everywhere on the domain $\Pi \subseteq \Sigma$, where

$$\Pi = \bigcup_{s>0} B_s$$

(why Π is connected?).

(ii) Let $k = 1, 2$. By (i), there is a domain $\Pi_k \subseteq \Omega_k$ such that

$$\mathrm{rng}(\tilde{\Gamma}) \subseteq \Pi_k \subseteq \Omega_k$$

and

$$u_k(x, y) = \frac{y}{x}$$

for all points $\begin{pmatrix} x \\ y \end{pmatrix} \in \Pi_k$. Consequently, the domain $\Pi = \Pi_1 \cap \Pi_2$ in $\mathbf{R}$ contains the range of the base curve $\tilde{\Gamma}$ and

$$u_1(x, y) = u_2(x, y)$$

for all $\begin{pmatrix} x \\ y \end{pmatrix} \in \Pi$. $\square$

To summarize the work done, we have found, with the use of Proposition 6.4.1, a solution of the Cauchy problem (6.4.3) on any domain Ω in $\mathbf{R}$ which does not meet the line $x = 0$ and contains the range of $\tilde{\Gamma}$, and used Theorem 6.2.1 to investigate the matter of uniqueness. In particular, it turned out that if $u_1 \in C^1(\Omega_1)$ and $u_2 \in C^1(\Omega_2)$ are solutions of the Cauchy problem (6.4.3), then the functions u_1 and u_2 coincide on an open and connected neighborhood of the base curve $\tilde{\Gamma}$. $\square$

Definition (Equivalence of solutions of a Cauchy problem for a first-order PDE). Consider a Cauchy problem

$$\begin{cases} \mathcal{E}(u) = 0, \\ u(f(s), g(s)) = h(s), \quad (s \in I) \end{cases} \tag{6.4.6}$$

for a first-order partial differential equation

$$\mathcal{E}(u) = 0$$

on an open set $\mathcal{O}$ in $\mathbf{R}^2$.

Solutions $u_1 \in C^1(\Omega_1)$ and $u_2 \in C^1(\Omega_2)$ of the Cauchy problem (6.4.6), where Ω_1, Ω_2 are suitable subdomains of $\mathcal{O}$, are said to be *equivalent*, symbolically,

$$u_1 \sim u_2,$$

if there is a domain Π in $\mathbf{R}^2$ which is contained in $\Omega_1 \cap \Omega_2$ and which contains the range $\mathrm{rng}(\tilde{\Gamma}) = \tilde{\Gamma}(I)$ of the base curve $\tilde{\Gamma}$ associated with the Cauchy problem (6.4.6),

$$\tilde{\Gamma}(I) \subseteq \Pi \subseteq \Omega_1 \cap \Omega_2,$$

such that

$$u_1(x, y) = u_2(x, y)$$

for all $\begin{pmatrix} x \\ y \end{pmatrix} \in \Pi$.

Claim 6.4.2. *The relation* ~ *is an equivalence relation on the set of solutions of the Cauchy problem (6.4.6).*

The straightforward proof is left to the reader.

Now we can say that the Cauchy problem (6.4.6) *has a unique solution up to equivalence*, or, for simplicity's (confusion's?) sake, *has a unique solution*, if the corresponding set of solutions has exactly one equivalence class by the relation ~. For instance, the Cauchy problem in Example 6.4.1 has, as we have seen above, a unique solution up to equivalence.

Similarly, we say that the Cauchy problem (6.4.6) has *infinitely many pairwise nonequivalent solutions*, or simply *infinitely many solutions*, if the set of solutions has infinitely many equivalence classes by the relation ~. Speaking of an example of the last situation, as well as an example of the situation when a given Cauchy problem has no solution, the reader may consult the *proof* of Claim 7.2.6 from the next chapter. △

Local solutions. Understandably, when dealing with a particular Cauchy problem, there may be a very slim, or no chance at all to obtain a global solution, but there may be good chances to obtain *local* solutions. On the intuitive level, the notion of a local solution of a Cauchy problem must be quite clear, but needs formalizing (for it is indeed not only proper, but also a *safer* approach).

Definition (Local solutions of Cauchy problems for first-order PDEs). Let

$$\begin{cases} \mathcal{E}(u) = 0, \\ u(f(s), g(s)) = h(s), & (s \in I) \end{cases} \tag{6.4.7}$$

be a Cauchy problem for a first-order partial differential equation $\mathcal{E}(u) = 0$ on an open set $\mathcal{O}$ in $\mathbf{R}^2$ and let s_0 be an inner point of the interval I.

A function $u \in C^1(\Omega)$ is a *local solution* of the Cauchy problem (6.4.7) at $s = s_0$ if there is a positive number $\alpha > 0$ such that the interval $(s_0 - \alpha, s_0 + \alpha)$ is contained in I and u is a solution of the Cauchy problem

$$\begin{cases} \mathcal{E}(u) = 0, \\ u(f(s), g(s)) = h(s), & (s_0 - \alpha < s < s_0 + \alpha). \end{cases} \quad \triangle$$

We again defer the corresponding definition of equivalence of local solutions to a later moment (but we invite the reader to think on the matter).

Now let us discuss an important condition which makes existence of a local solution to a Cauchy problem possible.

The reader is requested to consider the following as a continuation of our discussion before Proposition 6.4.1.

Suppose we have managed to place a part of the curve Γ restricted on an open interval about $s_0 \in \mathrm{dom}(\Gamma)$ into the level set

$$H(x, y, z) = 0$$

of some first integral H, whose gradient ∇H does not vanish on an open subset of Λ about the point $\Gamma(s_0)$.

In view of the implicit function theorem (and in view of Proposition 6.4.1), to ensure existence of a solution of the functional equation

$$H(x, y, u(x, y)) = 0$$

for a C^1-function u, we would like the condition

$$H_z \neq 0$$

to be true on some open neighborhood of the point $\Gamma(s_0)$. Let us see what we could do about that.

The fact that a part of Γ lies in the level set of H implies easily that

$$f'H_x + g'H_y + h'H_z = 0$$

on some open interval about s_0. Further, being a first integral, H satisfies

$$aH_x + bH_y + cH_z = 0.$$

Now *if H_z had been equal to 0 at $\Gamma(s_0)$*, we would have that

$$\begin{cases} f'H_x + g'H_y = 0, \\ aH_x + bH_y = 0, \end{cases}$$

at $\Gamma(s_0)$ and this would have led to a contradiction with nonvanishing of the gradient ∇H of H *provided that*

$$\begin{vmatrix} f' & g' \\ a & b \end{vmatrix} \neq 0$$

at $\Gamma(s_0)$. As the reader may have already guessed, we are going to use the above condition in our next result, which we are now ready to present.

Theorem 6.4.3. *Let Λ be an open set in $\mathbf{R}^3$, let $a, b, c \in C^1(\Lambda)$ satisfy $(a, b, c) \neq (0, 0, 0)$ everywhere on Λ, let a map $\Gamma : I \to \mathbf{R}^3$, determined by functions $f, g, h \in C^1(I)$, be a smooth curve on an open interval I in $\mathbf{R}$ lying entirely in Λ, and let s_0 be a point of I.*

Suppose that

(a) there is an open neighborhood $O_P \subseteq \Lambda$ of the point $P = \Gamma(s_0)$ such that the vector field $\mathcal{F}_{a_1, b_1, c_1}$ on O_P, where a_1 (resp., b_1, c_1) is the restriction of a (resp., b, c) on O_P, has two functionally independent first integrals F, G;

(b) the condition

$$\begin{vmatrix} f'(s_0) & g'(s_0) \\ a(\Gamma(s_0)) & b(\Gamma(s_0)) \end{vmatrix} \neq 0$$

called the transversality condition *holds for $s = s_0$; in other words,*

$$\begin{vmatrix} f'(s_0) & g'(s_0) \\ a(f(s_0), g(s_0), h(s_0)) & b(f(s_0), g(s_0), h(s_0)) \end{vmatrix} \neq 0.$$

Then the Cauchy problem

$$\begin{cases} a(x, y, u)u_x + b(x, y, u)u_y = c(x, y, u), \\ u(f(s), g(s)) = h(s), \quad (s \in I) \end{cases}$$

has a local solution at $s = s_0$.

Observe that if the *transversality* condition holds for a point $s_0 \in I$, then necessarily the tangent vector $\Gamma'(s_0)$ to Γ drawn at the point $\Gamma(s_0)$ and the field vector $\mathcal{F}_{a,b,c}(\Gamma(s_0))$ at the point $\Gamma(s_0)$ are not parallel:

$$\Gamma'(s_0) \nparallel \mathcal{F}_{a,b,c}(\Gamma(s_0)),$$

and are sort of "crosswise oriented," hence the term "transversality."

Proof of Theorem 6.4.3. First, observe that there is a positive number $\delta_1 > 0$ such that

$$\Gamma(s) \in O_P$$

for all $s \in (s_0 - \delta_1, s_0 + \delta_1)$. Indeed, being an open set, O_P contains an open ball U_ε of radius $\varepsilon > 0$ centered at $P = \Gamma(s_0)$. Now due to continuity of the functions f, g, h, there exists $\delta_1 > 0$ such that whenever

$$|s - s_0| < \delta_1,$$

where $s \in I$, then

$$|f(s) - f(s_0)| < \varepsilon/\sqrt{3},$$

$$|g(s) - g(s_0)| < \varepsilon/\sqrt{3},$$
$$|h(s) - h(s_0)| < \varepsilon/\sqrt{3},$$

and, in effect, the point $\Gamma(s)$ is in U_ε.

Consider the functions

$$\alpha(s) = F(\Gamma(s)) = F(f(s), g(s), h(s))$$

and

$$\beta(s) = G(\Gamma(s))$$

on $(s_0 - \delta_1, s_0 - \delta_1)$. We claim that one of the functions $\alpha(s), \beta(s)$ has a nonzero derivative at s_0. Indeed, suppose otherwise. Since

$$\alpha'(s) = F_x(\Gamma(s))f'(s) + F_y(\Gamma(s))g'(s) + F_z(\Gamma(s))h'(s)$$

and

$$\beta'(s) = G_x(\Gamma(s))f'(s) + G_y(\Gamma(s))g'(s) + G_z(\Gamma(s))h'(s)$$

on I, vanishing of both α' and β' at s_0 means that

$$(\nabla F)(\Gamma(s_0)) \perp \Gamma'(s_0)$$

and

$$(\nabla G)(\Gamma(s_0)) \perp \Gamma'(s_0),$$

where $\perp$ denotes the orthogonality relation.

Now since, by the conditions, the gradient vectors of F, G at $\Gamma(s_0)$ must be linearly independent, and, by Proposition 5.3.1, the field vector

$$\mathcal{F}_{a_1, b_1, c_1}(\Gamma(s_0)) = \mathcal{F}_{a,b,c}(\Gamma(s_0))$$

at $\Gamma(s_0)$ must be orthogonal to both of them,

$$aF_x + bF_y + cF_z = 0,$$
$$aG_x + bG_y + cG_z = 0,$$

we get that

$$\mathcal{F}_{a,b,c}(\Gamma(s_0)) \parallel \Gamma'(s_0). \tag{6.4.8}$$

However, by condition (b), the matrix

$$\begin{pmatrix} a(\Gamma(s_0)) & b(\Gamma(s_0)) & c(\Gamma(s_0)) \\ f'(s_0) & g'(s_0) & h'(s_0) \end{pmatrix}$$

is of rank two, and so (6.4.8) is impossible.

Suppose that, for instance,

$$\alpha'(s_0) \neq 0,$$

and so, by the Inverse Function Theorem, α is locally C^1-invertible near s_0, which implies that there is a closed interval $J = [s_0 - \delta_2, s_0 + \delta_2]$, where $0 < \delta_2 < \delta_1$, on which the corresponding restriction $\alpha|_J$ of α is invertible. Suppose that $K = \alpha(J)$, where K is a finite closed interval in $\mathbf{R}$. Now the real-valued function

$$\varphi_0 = \beta \circ (\alpha|_J)^{-1}$$

on K is such that

$$\varphi_0(\alpha(s)) = \beta(s)$$

for all $s \in (s_0 - \delta_2, s_0 + \delta_2)$.

Next, it is easy to see that there is a C^1-extension $\varphi \in C^1(\mathbf{R})$ of φ_0 to the whole of $\mathbf{R}$ (see Exercise 3 to this chapter), and so we can set

$$H(x, y, z) = G(x, y, z) - \varphi(F(x, y, z)),$$

on O_P, thereby getting a first integral of the vector field $\mathcal{F}_{a_1, b_1, c_1}$ on O_P. As we have seen on a similar occasion above,

$$\nabla H = \nabla G - \varphi'(F)\nabla F$$

everywhere on O_P, and hence ∇H never vanishes on O_P due to linear independence of the gradient vectors of F and G everywhere on O_P.

As, by the construction,

$$H(f(s), g(s), h(s)) = H(\Gamma(s))$$
$$= \varphi(F(\Gamma(s)) - G(\Gamma(s))) = 0$$

for all $s \in (s_0 - \delta_2, s_0 + \delta_2)$, we have that

$$f'(s)H_x + g'(s)H_y + h'(s)H_z = 0$$

and, simultaneously, again by Proposition 5.3.1,

$$aH_x + bH_y + cH_z = 0$$

at every point of the form $\Gamma(s)$, where $s \in (s_0 - \delta_2, s_0 + \delta_2)$. As we have seen above, this implies that $H_z(\Gamma(s_0)) \neq 0$, for otherwise, we would have that

$$H_x = H_y = 0 \Rightarrow \nabla H = \begin{pmatrix} H_x \\ H_y \\ H_z \end{pmatrix} = \begin{pmatrix} 0 \\ 0 \\ 0 \end{pmatrix}$$

at $\Gamma(s_0)$, because, by (a), the determinant of the system

$$\begin{cases} f'(s_0)H_x + g'(s_0)H_y = 0 \\ aH_x + bH_y = 0 \end{cases}$$

at $\Gamma(s_0)$ is nonzero.

Since $H_z(\Gamma(s_0)) = H_z(P) \neq 0$ and since H_z is continuous on O_P, there is an open ball $B_P \subseteq O_P$ centered at P (a convex set) such that $H_z \neq 0$ everywhere on B_P.

Now as

$$H(\Gamma(s_0)) = H(f(s_0), g(s_0), h(s_0)) = 0$$

and

$$H_z(f(s_0), g(s_0), h(s_0)) \neq 0,$$

we have, by the implicit function theorem on real-valued functions in three variables (Proposition A.1.3), that there is an open ball Ω in $\mathbf{R}^2$ centered at $\tilde{P} = \begin{pmatrix} f(s_0) \\ g(s_0) \end{pmatrix}$ such that there is a (unique) function $v \in C^1(\Omega)$, whose graph is contained in B_P, which satisfies the functional equation

$$H(x, y, u(x, y)) = 0$$

everywhere on Ω. Finally, we choose a positive $\delta < \delta_2$ so that

$$\tilde{\Gamma}(s) = \begin{pmatrix} f(s) \\ g(s) \end{pmatrix} \in \Omega \quad \text{and} \quad \Gamma(s) = \begin{pmatrix} f(s) \\ g(s) \\ h(s) \end{pmatrix} \in B_P$$

for all $s \in (s_0 - \delta, s_0 + \delta)$.

So all the conditions of Proposition 6.4.1 are met by the open and convex set B_P, by the corresponding restrictions of $\mathcal{F}_{a,b,c}$, F, and G on B_P, by the domain Ω in $\mathbf{R}^2$, by the restriction of the curve Γ on $(s_0 - \delta, s_0 + \delta)$, and by the function φ. Consequently, by the proposition, there is a solution $u \in C^1(\Omega)$ of the Cauchy problem

$$\begin{cases} a(x, y, u)u_x + b(x, y, u)u_y = c(x, y, u), \\ u(f(s), g(s)) = h(s), \quad (s_0 - \delta < s < s_0 + \delta) \end{cases}$$

on Ω, which completes the proof. $\qquad\square$

Example 6.4.2. Consider the PDE

$$x(y - u)u_x + y(u - x)u_y = (x - y)u$$

and the space curve

$$\Gamma(s) = \begin{pmatrix} f(s) \\ g(s) \\ h(s) \end{pmatrix} = \begin{pmatrix} s \\ s \\ 1/s \end{pmatrix}$$

on $I = (1, +\infty)$.

Find local solutions of the corresponding Cauchy problem

$$\begin{cases} x(y - u)u_x + y(u - x)u_y = u(x - y), \\ u(s, s) = 1/s, \quad (s > 1), \end{cases}$$

for each $s_0 > 1$ and consider the problem of extending them to appropriate domains in $\mathbf{R}^2$.

Solution. Having done so much in this chapter on the theoretical front, let us allow ourselves to use math software to speed up the solution.

An immediate task is to get two first integrals of the corresponding vector field, which can be usually derived, as it has been discussed above, from the local description of solutions of the PDE. Unfortunately, the corresponding command for Maple produces nothing. Luckily, $W\!|\alpha$ answers to the query

```
x*(y-u(x,y))*D[ u(x,y),x ] + y*(u(x,y)-x)*D[ u(x,y),y ] =(x-y)*u(x,y)
```

with

$$c_1(-xyu(x, y), u(x, y) + x + y) = 0,$$

which means that the functions

$$F(x, y, z) = -xyz \quad \text{and} \quad G(x, y, z) = x + y + z$$

form a required pair of first integrals.

The remaining part of the solution can be carried out with Maple. We suggest that the reader create a text file with the template that follows. When dealing with a similar problem, copy everything in the template file and paste to Maple (as a program).

Read the program, which follows the ideas in the proof of Theorem 6.4.3, and try figure out what is what in it.

```
            restart;
            with(linalg):

###         Coefficients:

            a := (x,y,z) -> ... :
            b := (x,y,z) -> ... :
            c := (x,y,z) -> ... :

###         Initial curve:

             f := s-> ... :
             g := s-> ... :
             h := s-> ... :

            Gamma := s-> (f(s), g(s),h(s)):

###         First integrals:

             F := (x,y,z) -> ... :
             G := (x,y,z) -> ... :

###         Gradients of first integrals,
###              for condition (a) in Theorem 4.3:

            jacobian( [F(x,y,z), G(x,y,z),0],
                        [x,y,z]  );

###         Transversality condition,
###              condition (b) in Theorem 4.3:

            DET:=det(
              matrix(2,2,[diff(f(s),s), diff(g(s),s),
              a(Gamma(s)), b(Gamma(s))])
              );

            alpha := s -> F(Gamma(s)):
            beta   := s -> G(Gamma(s)):
```

```
                print(alpha(s), beta(s));

#2#             phi := s -> ... ;

###             First integral whose level set
###                 contains the initial curve

#2#             H := (x,y,z) ->
#2#                     phi(G(x,y,z))-F(x,y,z) :

#2#             H := (x,y,z) ->
#2#                     phi(F(x,y,z))-G(x,y,z) :

#2#             H(x,y,z);

###             Solutions of the functional equation
###                 H(x,y,u(x,y))=0

#2#             solve( H(x,y,z), z);
```

Let us learn how to use the template above.

The first thing to do is to replace dots ... in UNCOMMENTED lines (those lines that do not begin with a # symbol) with relevant info:

- on the coefficient functions a, b, c;
- on the initial curve

$$\Gamma(s) = \begin{pmatrix} f(s) \\ g(s) \\ h(s) \end{pmatrix},$$

- and on the first integrals F, G.

Having Maple's response, we need to find a function $\varphi \in C^1(\mathbf{R})$ such that

$$\varphi(F(\Gamma(s))) = G(\Gamma(s)), \tag{6.4.9a}$$

or

$$\varphi(G(\Gamma(s))) = F(\Gamma(s)) \tag{6.4.9b}$$

everywhere on the domain of the initial curve. Once the defining formula for φ is found, enter it and uncomment the corresponding line. Next, depending on which of the formulas (6.4.9a), or (6.4.9b) is true for your function, uncomment the corresponding lines, defining the first integral H:

```
#2#        H := (x,y,z) ->
#2#                  phi(G(x,y,z))-F(x,y,z) :


#2#        H := (x,y,z) ->
#2#                  phi(F(x,y,z))-G(x,y,z) :
```

Having done that, remove all other symbols

```
#2#
```

below this part of the program, and press <Enter> to run the program again (for the second time).

Let us see how this works for the problem at hand. We enter available data first:

```
          restart;
          with(linalg):


###        Coefficients:


          a := (x,y,z) -> x*(y-z) :
          b := (x,y,z) -> y*(z-x) :
          c := (x,y,z) -> z*(x-y) :


###        Initial curve:


          f := s-> s    :
          g := s-> s    :
          h := s-> 1/s :


          Gamma := s-> (f(s), g(s),h(s)):


###        First integrals:


          F := (x,y,z) -> -x*y*z :
          G := (x,y,z) -> x+y+z   :


###        Gradients of first integrals ...


          jacobian( [F(x,y,z), G(x,y,z),0],
                    [x,y,z]);


###        Transversality condition ...


          DET:=det(
```

```
matrix(2,2,[diff(f(s),s), diff(g(s),s),
a(Gamma(s)), b(Gamma(s))])
);

alpha := s -> F(Gamma(s)):
beta  := s -> G(Gamma(s)):

print(alpha(s), beta(s));
```

Maple's response to the above code will be

$$\begin{pmatrix} -yz & -xz & -xy \\ 1 & 1 & 1 \\ 0 & 0 & 0 \end{pmatrix}$$

$$DET := 2s^2 - 2$$

$$-s, 2s + \frac{1}{s}$$

which should be then interpreted as follows:

1) After removing the zero row from the Jacobian matrix above, we get that

$$\begin{pmatrix} F_x & F_y & F_z \\ G_x & G_y & G_z \end{pmatrix} = \begin{pmatrix} -yz & -xz & -xy \\ 1 & 1 & 1 \end{pmatrix},$$

on $\mathbf{R}^3$, and hence for every $s > 1$,

$$\mathrm{rank}\begin{pmatrix} F_x & F_y & F_z \\ G_x & G_y & G_z \end{pmatrix}$$

$$= \mathrm{rank}\begin{pmatrix} -1 & -1 & -s^2 \\ 1 & 1 & 1 \end{pmatrix} = 2$$

at the corresponding point $\Gamma(s)$, which means that ∇F and ∇G are linearly independent along Γ (and it will be also true for suitable *open neighborhoods* of all points on Γ).

2) for all $s > 1$, we have that

$$\begin{vmatrix} a(\Gamma(s)) & b(\Gamma(s)) \\ f'(s) & g'(s) \end{vmatrix} = 2s^2 - 2,$$

and so the transversality condition, condition (b) in Theorem 6.4.3, is true everywhere on $I = (1, +\infty)$.

3) We have that

$$\alpha(s) = F(\Gamma(s)) = -s$$

and

$$\beta(s) = G(\Gamma(s)) = 2s + \frac{1}{s}$$

for all $s > 1$. An evident C^1-function to take $\alpha(s)$ to $\beta(s)$ is the function

$$\varphi(s) = -2s - \frac{1}{s}.$$

So by Theorem 6.4.3, there must be a local solution for every $s_0 > 1$.

We change accordingly the remaining lines in the template, thereby getting

```
phi := s -> -2*s-1/s;

###     First integral whose level set
###        contains the initial curve

#2#     H := (x,y,z) ->
#2#         phi(G(x,y,z))-F(x,y,z);

        H := (x,y,z) ->
            phi(F(x,y,z))-G(x,y,z);

        H(x,y,z);

###     Solutions of the functional equation
###        H(x,y,u(x,y))=0

        solve( H(x,y,z), z);
```

and run the program once again (finish editing, and then press <Enter>).

Maple will respond with the first integral $H = \varphi(F) - G$, given as

$$2xyz + \frac{1}{xyz} - x - y - z$$

and with two solutions of the equation

$$H(x, y, z) = 0$$

for z: given as

$$\frac{1}{2} \frac{x^2 y + xy^2 + \sqrt{x^4 y^2 + 2x^3 y^3 + x^2 y^4 - 8x^2 y^2 + 4xy}}{2x^2 y^2 - xy}$$

and

$$\frac{1}{2} \frac{x^2y + xy^2 - \sqrt{x^4y^2 + 2x^3y^3 + x^2y^4 - 8x^2y^2 + 4xy}}{2x^2y^2 - xy}.$$

Apparently, the above formulas are the defining formulas for two potential local solutions of the Cauchy problem at a point $s_0 > 1$, which we denote by $u_1(x, y)$ and $u_2(x, y)$, respectively. According to the theorem, one of them should be a local solution at the point s_0, namely, that one, which meets the initial condition.

Let $\delta > 0$ be any positive number such that $(s_0 - \delta, s_0 + \delta) \subseteq I$. We have that

$$u_1(s, s) = \frac{2s^3 + \sqrt{4s^6 - 8s^4 + 4s^2}}{2(2s^4 - s^2)},$$

whence, since for every $s \in (s_0 - \delta, s_0 + \delta)$, we have that $s > 1$,

$$u_1(s, s) = \frac{s^3 + |s^3 - s^2|}{s(2s^3 - s^2)} = \frac{1}{s}$$

for all $s \in (s_0 - \delta, s_0 + \delta)$. Incidentally, this shows that the function $u_1(x, y)$ exists on an open neighborhood of the curve $\tilde{\Gamma}(s) = \begin{pmatrix} s \\ s \end{pmatrix}$ on $(1, +\infty)$. As for the function u_2, we have that

$$u_2(s, s) = \frac{s^3 - s^3 + s^2}{2s^4 - s^2} = \frac{1}{2s^2 - 1} \not\equiv \frac{1}{s}$$

on $(s_0 - \delta, s_0 + \delta)$.

We then conclude that the function

$$u_1(x, y) = \frac{1}{2} \frac{x^2y + xy^2 + \sqrt{x^4y^2 + 2x^3y^3 + x^2y^4 - 8x^2y^2 + 4xy}}{2x^2y^2 - xy}$$

defined on an appropriate open disk centered at $\begin{pmatrix} s_0 \\ s_0 \end{pmatrix}$ is a local solution of the Cauchy problem under consideration at $s_0 > 1$. (To quell any remaining doubt, if it still exists, verify using Maple, that u_1 does satisfy the PDE in question).

To summarize, given any domain Ω in $\mathbf{R}^2$ which contains the curve $\tilde{\Gamma}$ on $(1, +\infty)$ and for which the defining formula above,

$$\frac{1}{2} \frac{x^2y + xy^2 + \sqrt{x^4y^2 + 2x^3y^3 + x^2y^4 - 8x^2y^2 + 4xy}}{2x^2y^2 - xy}$$

exists for all $\begin{pmatrix} x \\ y \end{pmatrix} \in \Omega$, the function

$$v(x, y) = \frac{1}{2} \frac{x^2y + xy^2 + \sqrt{x^4y^2 + 2x^3y^3 + x^2y^4 - 8x^2y^2 + 4xy}}{2x^2y^2 - xy}$$

on Ω, is a solution of the corresponding Cauchy problem.

Finally, a couple of practical advices. To make verification of the initial condition with Maple: create a necessary number of commands like

```
u1 := (x,y) -> defining formula 1 :
u2 := (x,y) -> defining formula 2 :
```

. . .

and then add verifications of the initial condition, say in the form

```
print(`1st function: `,
          u1( f(s),g(s) ) - h(s) );

print(`2nd function: `,
          u2( f(s),g(s) ) - h(s) );
```

. . .

to your program. The defining formulas can be copied one by one from the output presenting the solutions of the equation $H(x,y,z) = 0$ for z.

A small trick: if it is not immediately evident for which of the potential solutions $u = u_i$ we have that

$$u(f(s),g(s)) - h(s) = 0$$

everywhere about a point s_0 in question, draw the graphs of all the functions

$$u(f(s),g(s)) - h(s)$$

on a suitable interval about s_0, using commands like

```
plot( u(f(s),g(s))-h(s), s = s0 - 0.01 .. s0 +0.01 );
```

A potential solution u for which the graph of the function $u(f(s),g(s))-h(s)$ is extremely close to the x-axis (say, lying between the lines $y = \pm 10^{-10}$) must be, in all likelihood, the required solution. □

Exercises

1. For each of the of the following first-order linear PDEs:
(a) consider the equation as a quasilinear one, and write down the corresponding characteristic system;
(b) determine whether it has two functionally independent first integrals on the specified open set Λ in $\mathbf{R}^3$;

(c) if yes, obtain a description of solutions of the PDE on the specified domain Ω in $\mathbf{R}^2$; finally, confer your result with the corresponding result considered above

(as a matter of fact, you should have performed steps (a,b) when working on Exercise 5.3, and so please review your solution carefully; a fresh start would be even better):

(i) $u_x = 0$, $\Lambda = \mathbf{R}^3$, $\Omega = $ any domain in $\mathbf{R}^2$ (cf. Corollary 1.4.2);

(ii) $xu_x + yu_y = u$, $\Lambda = \mathbf{R}^* \times \mathbf{R} \times \mathbf{R}$, $\Omega = \mathbf{R}^+ \times \mathbf{R}$ (cf. Example 2.1.1);

(iii) $xu_x - yu_y = 2x^2$, $\Lambda = \mathbf{R}^* \times \mathbf{R} \times \mathbf{R}$, $\Omega = \mathbf{R}^+ \times \mathbf{R}^+$ (cf. Example 2.3.1);

(iv) $yu_x - xu_y = 0$, $\Lambda = \mathbf{R}^* \times \mathbf{R} \times \mathbf{R}$, $\Omega = \mathbf{R}^+ \times \mathbf{R}$ (cf. Exercise 2.1),

where $\mathbf{R}^*$ denotes the set $\mathbf{R} \setminus \{0\}$ of all nonzero real numbers.

2. For each of the following first-order quasilinear PDEs:

(a) write down the corresponding characteristic system and find two first integrals;

(b) find an open set Λ in $\mathbf{R}^3$ on which the first integrals are functionally independent;

(c) describe solutions $u \in C^1(\Omega)$ of the PDE whose graph is contained in Λ for domains Ω in $\mathbf{R}^2$ satisfying

$$\Omega \subseteq \mathrm{proj}_{x,y}(\Lambda)$$

(again, no harm in doing steps (a,b) for the second time, especially if you had difficulties earlier):

(i) $uu_x + yu_y = x$;

(ii) $uu_x - yu_y = x$;

(iii) $u_x + uu_y = -u^2$;

(iv) $u_x + uu_y = u$;

(v) $x(y - u)u_x + y(x + u)u_y = (x + y)u$.

3. Consider a continuous real-valued function $f : [a, b] \to \mathbf{R}$ on a finite closed interval in $\mathbf{R}$. Suppose that f is continuously differentiable on (a, b) and that the limits

$$\lim_{x \to a_+} f'(x) \quad \text{and} \quad \lim_{x \to b_-} f'(x)$$

exist. Show then that there is a continuously differentiable function $\varphi \in C^1(\mathbf{R})$ such that

$$\varphi(x) = f(x)$$

for all $x \in [a, b]$.

4. Use Theorem 6.4.3 and the program template from Example 6.4.2 to analyze the local solutions of the following Cauchy problems, and then consider the problem of extending the local solutions to appropriate domains in $\mathbf{R}^2$:

(i) $\begin{cases} u_x + uu_y = 0, \\ u(s, 1) = s + 1, \quad (s > -1), \end{cases}$

(ii) $\begin{cases} u_x + uu_y = 0, \\ u(s, s) = s, \quad (s > 1), \end{cases}$

(iii) $\begin{cases} (y + u)u_x + yu_y = x - y, \quad (y \neq 0), \\ u(-s, -s) = s, \quad (s < 0), \end{cases}$

(iv) $\begin{cases} uu_x + yu_y = x, \quad (y \neq 0), \\ u(s, 1) = 1 + s, \quad (s > 0), \end{cases}$

(v) $\begin{cases} u_x + uu_y = u, \\ u(0, s) = s + 1, \quad (s \in \mathbf{R}), \end{cases}$

(vi) $\begin{cases} (y + u)u_x + yu_y = x - y, \quad (y \neq 0), \\ u(s, 1) = s + 1, \quad (s > 0). \end{cases}$

7 Method of characteristics for first-order quasilinear equations

The last result in the previous chapter, Theorem 6.4.3, enables one to obtain *local* solutions of Cauchy problems involving first-order quasilinear PDEs. However, it is only too easy to point out the obvious weaknesses of Theorem 6.4.3:

- first of all, it requires two functionally independent first integrals of the corresponding vector field (and finding those is more an "art" than a "science");
- the theorem only states existence of a domain in $\mathbf{R}^2$ to consider the corresponding PDE on, but provides no way of constructing such a domain explicitly;
- and so on.

The method of obtaining of solutions of Cauchy problems for first-order quasilinear equations we shall study in this chapter gives the user rather more control and, importantly, is highly attractive from an aesthetic point of view.

The key idea of the method belongs to Gaspard Monge (1746–1818) who was (one of) the first to realize that ("normally")

> the graph of any solution of a first-order quasilinear PDE is "made" up of its characteristic curves, and vice versa.

We shall see below that the rigorous proof of this statement requires the full force of the Peano–Pickard–Lindelöf theorem, all whose contributors were born after Monge's demise, and so it is safer to call the statement *Monge's hypothesis*, or *Monge's postulate*.

7.1 Surfaces made up of characteristic curves

Let us first do the sufficiency part of Monge's postulate ("vice versa"), an easier one.

Proposition 7.1.1. *Let Λ be an open set in $\mathbf{R}^3$, let functions $a, b, c \in C^1(\Lambda)$ be continuously differentiable on Λ, and let Ω be a domain in $\mathbf{R}^2$ such that $\Omega \subseteq \mathrm{proj}_{x,y}(\Lambda)$. Suppose that the graph*

$$
S = \left\{ \begin{pmatrix} x \\ y \\ z \end{pmatrix} \in \mathbf{R}^3 : z = u(x,y), \begin{pmatrix} x \\ y \end{pmatrix} \in \Omega \right\}
$$

of a function $u \in C^1(\Omega)$ is a union of the characteristic curves of the vector field $\mathcal{F}_{a,b,c}$ on Λ. Then u is a solution of the quasilinear PDE

$$
a(x,y,u)u_x + b(x,y,u)u_y = c(x,y,u)
$$

on Ω.

https://doi.org/10.1515/9783110677256-007

Proof. Let us decipher what "a union of the of the characteristic curves" means.

Take an arbitrary point $P = \begin{pmatrix} x_0 \\ y_0 \\ z_0 \end{pmatrix}$ of S. As $P \in S$, then

$$z_0 = u(x_0, y_0).$$

Now S being a union of characteristic curves means that there is a characteristic curve

$$\gamma(t) = \begin{pmatrix} x(t) \\ y(t) \\ z(t) \end{pmatrix}, \quad (t \in I),$$

where I is an open interval in $\mathbf{R}$, which passes through P at a certain $t_0 \in I$:

$$\gamma(t_0) = \begin{pmatrix} x_0 \\ y_0 \\ z_0 \end{pmatrix} \iff \begin{cases} x_0 = x(t_0), \\ y_0 = y(t_0), \\ z_0 = z(t_0), \end{cases}$$

and which is *contained* in S, that is,

$$\gamma(t) \in S \quad \text{for all } t \in I.$$

Importantly, since a point $\begin{pmatrix} x \\ y \\ z \end{pmatrix}$, where $\begin{pmatrix} x \\ y \end{pmatrix} \in \Omega$, belongs to S if and only if $z = u(x, y)$, we get that

$$z(t) = u\big(x(t), y(t)\big) \tag{7.1.1}$$

for all t in I. Finally, as γ is a characteristic curve, we have that

$$\begin{cases} x'(t) = a(x(t), y(t), z(t)), \\ y'(t) = b(x(t), y(t), z(t)), \quad (t \in I). \\ z'(t) = c(x(t), y(t), z(t)) \end{cases}$$

Now let us differentiate both parts of our "important" equation

$$z(t) = u\big(x(t), y(t)\big)$$

by t. On one hand,

$$z'(t) = c\big(x(t), y(t), z(t)\big),$$

because y is a characteristic curve. On the other hand,

$$[u(x(t), y(t))]'$$
$$= u_x(x(t), y(t))x'(t) + u_y(x(t), y(t))y'(t),$$

and again, since y is a characteristic curve,

$$[u(x(t), y(t))]'$$
$$= u_x(x(t), y(t)) \underbrace{a(x(t), y(t), z(t))}_{x'(t)}$$
$$+ u_y(x(t), y(t)) \underbrace{b(x(t), y(t), z(t))}_{y'(t)}.$$

Therefore,

$$c(x(t), y(t), z(t))$$
$$= u_x(x(t), y(t))a(x(t), y(t), z(t))$$
$$+ u_y(x(t), y(t))b(x(t), y(t), z(t)),$$

for all points $t \in I$.

In particular, for $t = t_0$ we have that

$$c(x_0, y_0, z_0)$$
$$= u_x(x_0, y_0)a(x_0, y_0, z_0) + u_y(x_0, y_0)b(x_0, y_0, z_0),$$

or, since $z_0 = u(x_0, y_0)$,

$$c(x_0, y_0, u(x_0, y_0))$$
$$= u_x(x_0, y_0)a(x_0, y_0, u(x_0, y_0))$$
$$+ u_y(x_0, y_0)b(x_0, y_0, u(x_0, y_0)).$$

As $\begin{pmatrix} x_0 \\ y_0 \end{pmatrix}$ is an *arbitrary* point of Ω, we obtain that

$$c(x, y, u) = a(x, y, u)u_x + b(x, y, u)u_y,$$

and the function $u = u(x, y)$ is indeed a solution of our equation. □

The method of solving Cauchy problems for quasilinear equations developed by Gaspard Monge, who is considered as the father of differential geometry, can be briefly summarized as follows.

Suppose we are given an open set Λ in $\mathbf{R}^3$, functions $a, b, c \in C^1(\Lambda)$ that are continuously differentiable on Λ, and a space curve $\Gamma : I \to \mathbf{R}^3$, where

$$\Gamma(s) = \begin{pmatrix} f(s) \\ g(s) \\ h(s) \end{pmatrix}, \quad (s \in I),$$

which lies entirely in Λ. Let further Ω be any domain in $\mathbf{R}^2$ such that

$$\Omega \subseteq \mathrm{proj}_{x,y}(\Lambda)$$

and such that the orthogonal projection $\widetilde{\Gamma}$ of Γ on the xy-plane, the plane curve

$$\widetilde{\Gamma}(s) = \begin{pmatrix} f(s) \\ g(s) \end{pmatrix}, \quad (s \in I)$$

lies entirely in Ω. Now it makes sense to consider the Cauchy problem requiring to find a solution $u \in C^1(\Omega)$ of the quasilinear PDE

$$a(x, y, u)u_x + b(x, y, u)u_y = c(x, y, u) \tag{7.1.2}$$

on Ω, whose graph contains (the range of) Γ, that is,

$$u\big(f(s), g(s)\big) = h(s)$$

for all $s \in I$.

Having done a lot of work on Cauchy problems in the previous chapter, the reader may well agree that what matters most in the above setup is the open set Λ, the coefficient functions a, b, c, and the space curve Γ. It is "fair, just and reasonable" not to be too specific as far the choice of the domain Ω is concerned. For, "normally," it is enough to be able to prove that there is at least one such a domain for which the corresponding Cauchy problem has a solution.

Then in order to obtain a solution to the above problem, we can, according to Monge, proceed as follows (warning: no guarantee in getting a solution at the end).

Step 1. Draw the initial curve

$$\Gamma(s) = \begin{pmatrix} f(s) \\ g(s) \\ h(s) \end{pmatrix},$$

where s runs over I.

Step 2. Fix some real number t_0 (say, $t_0 = 0$), and for every $s \in I$, draw a characteristic curve $\gamma_s : J_s \to \mathbf{R}^3$ of the vector field $\mathcal{F}_{a,b,c}$ whose domain is an open interval J_s in $\mathbf{R}$ about t_0, which passes through the corresponding point $\begin{pmatrix} f(s) \\ g(s) \\ h(s) \end{pmatrix}$ on Γ at $t = t_0$,

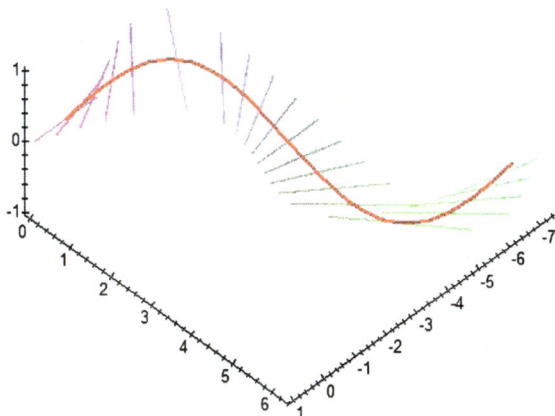

Figure 7.1.1: Monge Surface: formation.

that is, $\gamma_s(t_0) = \begin{pmatrix} f(s) \\ g(s) \\ h(s) \end{pmatrix}$, and which lies entirely in Λ (say, in the case when $t_0 = 0$, we

can choose J_s to be an open interval of the form $(-\varepsilon_s, \varepsilon_s)$, where $\varepsilon_s > 0$). We also take care to make sure that the set

$$\Theta = \bigcup_{s \in I} \{s\} \times J_s$$

is a domain in $\mathbf{R}^2$ (of the st-plane).

Figure 7.1.1 shows the process of formation of the Monge surface associated with a Cauchy problem involving Burgers' equation (the resulting surface is shown in Figure 7.1.2):

- an initial space curve Γ is drawn in red;
- a number of characteristic curves of Burgers' equation (the straight lines in the figure) are drawn, all of them coming through the corresponding points on Γ at $t_0 = 0$, as is explained above.

Step 3. Taking the union of (the ranges of) the curves γ_s, we get a *two-parametric surface* which will be referred to as the *Monge surface*, or as the *characteristic surface*, associated with the Cauchy problem at hand:

$$\mathcal{M} = \{\gamma_s(t) : s \in I, t \in J_s\}.$$

Why "*two-parametric?*" Well, every point P on $\mathcal{M}$ is determined by two parameters: by some $s \in I$ and by some $t \in J_s$ such that

$$P = \gamma_s(t).$$

Step 4. *If* the Monge surface will turned out to be the *graph* of some C^1-function on a domain Ω which meets our requirements, this function will be a solution to the

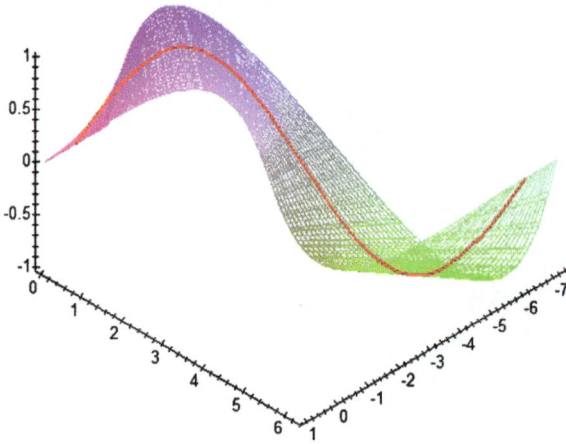

Figure 7.1.2: Monge surface.

corresponding Cauchy problem, for the graph of u certainly contains the initial curve Γ and the function u satisfies the PDE

$$a(x, y, u)u_x + b(x, y, u)u_y = c(x, y, u)$$

on Ω, since the Monge surface is "made" up of the characteristic curves of the corresponding vector field $\mathcal{F}_{a,b,c}$ (Proposition 7.1.1).

To refresh the reader's memory about two-parametric surfaces, let us recall the corresponding definition and consider a couple of examples.

Very much like a space curve is defined via parametric equations

$$\gamma(t) = \begin{pmatrix} x(t) \\ y(t) \\ z(t) \end{pmatrix} \iff \begin{cases} x = f(t), \\ y = g(t), \\ z = h(t), \end{cases}$$

where t runs over some interval $I, \ldots$

$\ldots$ quite often, a *surface* Σ in $\mathbf{R}^3$ is defined via two-parametric equations

$$\begin{cases} x = X(s, t), \\ y = Y(s, t), \\ z = Z(s, t), \end{cases}$$

where a point $\begin{pmatrix} s \\ t \end{pmatrix}$ runs over some subset Θ of $\mathbf{R}^2$:

$$\Sigma = \left\{ \begin{pmatrix} X(s, t) \\ Y(s, t) \\ Z(s, t) \end{pmatrix} : \begin{pmatrix} s \\ t \end{pmatrix} \in \Theta \right\}.$$

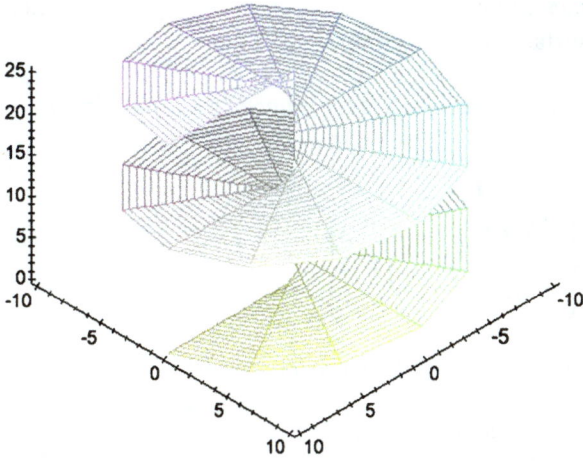

Figure 7.1.3: Helicoid.

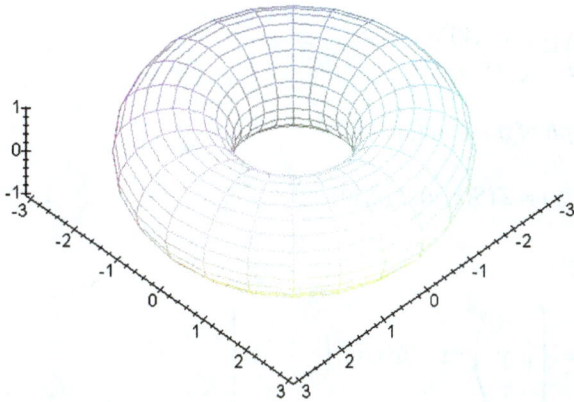

Figure 7.1.4: Torus.

Let use consider some examples.

Figure 7.1.3: a helicoid, a two-parametric surface in parameters r, φ:

$$\begin{cases} x = r\cos(\varphi), \\ y = r\sin(\varphi), \\ z = h\varphi, \end{cases} \quad \begin{pmatrix} r \\ \varphi \end{pmatrix} \in \mathbf{R}^{\geq 0} \times \mathbf{R}.$$

Figure 7.1.4: a *torus*, a two-parametric surface in parameters φ, ψ:

$$\begin{cases} x = (b + a\cos(\psi))\cos(\varphi), \\ y = (b + a\cos(\psi))\sin(\varphi), \\ z = a\sin(\psi), \end{cases} \quad \begin{pmatrix} \varphi \\ \psi \end{pmatrix} \in \mathbf{R} \times \mathbf{R}.$$

Now, with the above examples in mind, we see that an evident problem when working with a two-parametric surfaces is to determine

whether or not a given two-parametric surface Σ is the graph of some function u.

Proposition 7.1.2. *Let*

$$\Sigma = \left\{ \begin{pmatrix} X(s,t) \\ Y(s,t) \\ Z(s,t) \end{pmatrix} : \begin{pmatrix} s \\ t \end{pmatrix} \in \Theta \right\},$$

where $\Theta \subseteq \mathbf{R}^2$, be a two-parametric surface in $\mathbf{R}^3$. Suppose that the map

$$\Psi \begin{pmatrix} s \\ t \end{pmatrix} = \begin{pmatrix} X(s,t) \\ Y(s,t) \end{pmatrix}, \qquad \begin{pmatrix} s \\ t \end{pmatrix} \in \Theta$$

on Θ is invertible and that

$$\Psi^{-1} \begin{pmatrix} x \\ y \end{pmatrix} = \begin{pmatrix} S(x,y) \\ T(x,y) \end{pmatrix}, \qquad \begin{pmatrix} x \\ y \end{pmatrix} \in \Omega,$$

where $\Omega = \Psi(\Theta)$. Then Σ is the graph of the function

$$u(x,y) = Z\big(S(x,y), T(x,y)\big)$$

on Ω, that is,

$$\Sigma = \mathrm{Graph}(u) = \left\{ \begin{pmatrix} x \\ y \\ z \end{pmatrix} : z = u(x,y), \begin{pmatrix} x \\ y \end{pmatrix} \in \Omega \right\}.$$

Proof. Let $\begin{pmatrix} x \\ y \\ z \end{pmatrix}$ be a point of Σ. Then there is a point $\begin{pmatrix} s \\ t \end{pmatrix} \in \Theta$ such that

$$\begin{cases} x = X(s,t), \\ y = Y(s,t), \\ z = Z(s,t). \end{cases}$$

Accordingly, $\begin{pmatrix} x \\ y \end{pmatrix} = \Psi \begin{pmatrix} s \\ t \end{pmatrix}$ and then

$$\begin{pmatrix} s \\ t \end{pmatrix} = \Psi^{-1} \begin{pmatrix} x \\ y \end{pmatrix} = \begin{pmatrix} S(x,y) \\ T(x,y) \end{pmatrix}.$$

It follows that

$$z = Z(s,t) = Z(S(x,y), T(x,y)) = u(x,y).$$

Therefore,

$$\Sigma \subseteq \text{Graph}(u) = \left\{ \begin{pmatrix} x \\ y \\ z \end{pmatrix} : z = u(x,y), \begin{pmatrix} x \\ y \end{pmatrix} \in \Omega \right\}.$$

Conversely, suppose that a point $\begin{pmatrix} x \\ y \\ z \end{pmatrix}$ is in the graph of u. This means that

$\begin{pmatrix} x \\ y \end{pmatrix} \in \Omega$ and

$$z = u(x,y) = Z(S(x,y), T(x,y)) = Z \circ \Psi^{-1} \begin{pmatrix} x \\ y \end{pmatrix}. \tag{7.1.3}$$

Due to $\begin{pmatrix} x \\ y \end{pmatrix} \in \Omega = \Psi(\Theta)$, there is a uniquely determined point $\begin{pmatrix} s \\ t \end{pmatrix} \in \Theta$ such that

$$\begin{pmatrix} x \\ y \end{pmatrix} = \Psi \begin{pmatrix} s \\ t \end{pmatrix} = \begin{pmatrix} X(s,t) \\ Y(s,t) \end{pmatrix} \Rightarrow \begin{cases} x = X(s,t), \\ y = Y(s,t). \end{cases}$$

Then we deduce from (7.1.3) that

$$z = Z \circ \Psi^{-1} \circ \Psi \begin{pmatrix} s \\ t \end{pmatrix} = Z(s,t).$$

Therefore,

$$\begin{cases} x = X(s,t), \\ y = Y(s,t), \\ z = Z(s,t), \end{cases}$$

and so the point $\begin{pmatrix} x \\ y \\ z \end{pmatrix}$ is in Σ. Hence

$$\text{Graph}(u) \subseteq \Sigma,$$

and we conclude that

$$\text{Graph}(u) = \Sigma,$$

which completes the proof. □

As we have seen in Chapter 2, invertibility of the map Ψ above,

$$\Psi\begin{pmatrix} s \\ t \end{pmatrix} = \begin{pmatrix} X(s,t) \\ Y(s,t) \end{pmatrix}, \quad \begin{pmatrix} s \\ t \end{pmatrix} \in \Theta$$

means, in plain English, that it is possible to solve the system

$$\begin{cases} x = X(s,t), \\ y = Y(s,t), \end{cases}$$

for s and t, that is, to express s, t via x and y:

$$\begin{cases} x = X(s,t), \\ y = Y(s,t) \end{cases} \Rightarrow \begin{cases} s = S(x,y), \\ t = T(x,y), \end{cases}$$

and to show that, conversely,

$$\begin{cases} s = S(x,y), \\ t = T(x,y), \end{cases} \Rightarrow \begin{cases} x = X(s,t), \\ y = Y(s,t). \end{cases}$$

Surely, Monge's algorithm we have outlined above needs formalization, but, on the other hand, enough has been said to try to solve a simple Cauchy problem.

Example 7.1.1. Solve Burgers' equation

$$u_x + u u_y = 0$$

with the initial condition

$$u(0, 2s) = 3s, \quad (0 < s < 1).$$

Solution. According to Example 5.2.1, the characteristic curve of Burgers' equation passing through a point $\begin{pmatrix} x_0 \\ y_0 \\ z_0 \end{pmatrix}$ of $\mathbf{R}^3$ at $t = t_0 = 0$ is

$$\gamma(t) = \begin{cases} x(t) = t + x_0, \\ y(t) = z_0 t + y_0, \\ z(t) = z_0. \end{cases} \tag{7.1.4}$$

Let $s \in I = (0,1)$. Doing Step 2 of Monge's algorithm: we need to find a characteristic curve γ_s of Burgers' equation on a suitable open interval J_s about t_0 which passes through the point $P = \begin{pmatrix} 0 \\ 2s \\ 3s \end{pmatrix}$ at $t = t_0 = 0$. Using (7.1.4), we choose γ_s as follows:

$$\gamma_s(t) = \begin{pmatrix} t \\ 3st + 2s \\ 3s \end{pmatrix}, \quad t \in (-2/3, +\infty).$$

So, in the above notation, we set the interval

$$J_s = (-2/3, +\infty)$$

about $t_0 = 0$ to be the domain of y_s (it shall be clear soon why we choose $-2/3$ to be the left endpoint of J_s).

The resulting domain Θ of the st-plane is therefore

$$\Theta = \bigcup_{s \in I} \{s\} \times J_s = (0, 1) \times (-2/3, +\infty),$$

(the domain of the corresponding Monge surface, as a two-parametric surface – formally, a map on Θ taking values in $\mathbf{R}^3$ – to be considered at the next step).

Doing Step 3 of the algorithm: the Monge surface $\mathcal{M}$ is therefore

$$\mathcal{M} = \{y_s(t) : s \in (0, 1), t > -2/3\}$$

$$= \left\{ \begin{pmatrix} t \\ 3st + 2s \\ 3s \end{pmatrix} : s \in (0, 1), t > -2/3 \right\}$$

and, to repeat, it is the two-parametric surface determined by the functions

$$X(s, t) = t,$$
$$Y(s, t) = 3st + 2s,$$
$$Z(s, t) = 3s,$$

on

$$\Theta = (0, 1) \times (-2/3, +\infty);$$

in other words, the Monge surface is described by the following parametric equations:

$$\begin{cases} x = t, \\ y = 3st + 2s, \\ z = 3s, \end{cases} \quad \begin{pmatrix} s \\ t \end{pmatrix} \in \Theta. \tag{7.1.5}$$

By Proposition 7.1.2, the Monge surface is the graph of a function on a suitable subset of $\mathbf{R}^2$, if it is possible to solve the system, formed of the first two equations, for s and t.

Then, having obtained expressions for s, t in terms of x, y, we replace s, t in the formula for z, thereby getting a desired solution (again, by Proposition 7.1.2).

Let us try to realize this plan, which is tantamount to doing Step 4 of the algorithm. The system in question, formed of the first two equations of the Monge surface, is

$$\begin{cases} x = t, \\ y = 3st + 2s, \end{cases}$$

and we already have the expression for t via x:

$$t = x.$$

We further have that

$$\begin{cases} x = t, \\ y = 3st + 2s, \end{cases} \Rightarrow y = s(3t + 2) = s(3x + 2)$$

$$\Rightarrow s = \frac{y}{3x + 2},$$

provided that $x \neq -2/3$. Conversely, whenever $x \neq -2/3$, we have, as it is easy to see, that

$$\begin{cases} t = x, \\ s = \dfrac{y}{3x + 2} \end{cases} \Rightarrow \begin{cases} x = t, \\ y = 3st + 2s. \end{cases}$$

Defining the function Ψ on $\Theta = (0, 1) \times (-2/3, +\infty)$ by

$$\Psi\begin{pmatrix} s \\ t \end{pmatrix} = \begin{pmatrix} t \\ 3st + 2s \end{pmatrix}, \quad \begin{pmatrix} s \\ t \end{pmatrix} \in \Theta,$$

we quickly obtain that

$$\Omega = \Psi(\Theta) = \left\{ \begin{pmatrix} x \\ y \end{pmatrix} : -2/3 < x,\ 0 < y < 3x + 2 \right\},$$

since

$$\begin{cases} 0 < s < 1, \\ -2/3 < t \end{cases} \quad \Longleftrightarrow \quad \begin{cases} -2/3 < t, \\ 0 < 3st + 2s < 3t + 2 \end{cases}$$

(proof?). Observe also that Ω contains the (range of the) orthogonal projection

$$\tilde{\Gamma}(s) = \begin{pmatrix} 0 \\ 2s \end{pmatrix}, \quad (0 < s < 1)$$

of the initial curve Γ (as it should; why?). We recommend the reader to sketch Θ in the st-plane, and Ω and $\tilde{\Gamma}$ in the xy-plane.

Using the third equation in (7.1.5), we then get that

$$z = 3s = \frac{3y}{3x + 2},$$

or, formally, that the Monge surface (7.1.5) is in fact, by Proposition 7.1.2, the graph of the function

$$u(x, y) = /\!/ \quad 3s \quad /\!/ = \frac{3y}{3x + 2}.$$

on Ω. According to Monge's argument above, this function satisfies Burgers' equation on Ω (by Proposition 7.1.1, since it is evidently C^1 on Ω and its graph is "made" up of characteristic curves) and contains the initial curve

$$\Gamma(s) = \begin{pmatrix} 0 \\ 2s \\ 3s \end{pmatrix}$$

on $I = (0,1)$, and hence provides a solution to our Cauchy problem.

Suppose, however, that one is unconvinced either by Monge's argument, or by the proof of Proposition 7.1.1, or by the proof of Proposition 7.1.2.

Let us then verify that the function

$$u(x,y) = \frac{3y}{3x+2}$$

does satisfy Burgers' equation on Ω and the initial condition saying that

$$u(0,2s) = 3s$$

for all points $s \in (0,1)$.

Clearly,

$$\left[\frac{3y}{3x+2}\right]_x + \frac{3y}{3x+2}\left[\frac{3y}{3x+2}\right]_y =$$
$$-\frac{9y}{(3x+2)^2} + \frac{3y}{3x+2}\cdot\frac{3}{3x+2} =$$
$$0$$

everywhere on Ω and

$$u(0,2s) = \frac{3\cdot 2s}{2} = 3s$$

for all $s \in (0,1)$.

As we have mentioned above, when solving a particular Cauchy problem, the matter of choosing the domain in $\mathbf{R}^2$ to consider the corresponding PDE on is not that important (unless otherwise required). In most of cases, an answer stating that there is a domain Ω on which the corresponding PDE has a solution meeting the corresponding initial condition will be deemed quite satisfactory. On these grounds, we could have skipped determination of the image $\Psi(\Theta)$ above, once it has become clear that Ψ is C^1-invertible, and then conclude that given any domain Ω in $\mathbf{R}^2$ which does not meet the line $x = -2/3$ and which contains the range of the projection $\bar{\Gamma}$ of Γ, the function

$$u(x,y) = \frac{3y}{3x+2}$$

is a solution of the corresponding Cauchy problem. $\square$

7.2 Main theorem

Next, we prove the necessity part of Monge's postulate, that is, prove that, "normally," the graph of any solution of a first-order quasilinear equation is "made" up of characteristic curves.

Proposition 7.2.1. *Let Λ be an open set in $\mathbf{R}^3$, let $a, b, c \in C^1(\Lambda)$, and let Ω be a domain in $\mathbf{R}^2$ such that $\Omega \subseteq \mathrm{proj}_{x,y}(\Lambda)$.*
Suppose that a function $u \in C^1(\Omega)$ is a solution of the PDE

$$a(x, y, u)u_x + b(x, y, u)u_y = c(x, y, u) \tag{7.2.1}$$

on Ω whose graph is contained in Λ, and that t_0 is any real number. Then:
(i) The graph of u is a union of the characteristic curves of the PDE (7.2.1) (of the vector field $\mathcal{F}_{a,b,c}$ on Λ).

(ii) Equivalently, given any point $\begin{pmatrix} x_0 \\ y_0 \end{pmatrix} \in \Omega$, there is a solution of the system

$$\begin{cases} x'(t) = a(x(t), y(t), u(x(t), y(t))), \\ y'(t) = b(x(t), y(t), u(x(t), y(t))) \end{cases}$$

of ODEs on an open interval I about t_0 such that $\begin{pmatrix} x(t) \\ y(t) \end{pmatrix} \in \Omega$ for all $t \in I$, and

$$x(t_0) = x_0 \quad and \quad y(t_0) = y_0$$

at $t = t_0$.
(iii) Suppose $u_1, u_2 \in C^1(\Omega)$ are solutions of the PDE (7.2.1), whose graphs S_1, S_2 are both contained in Λ and have nonempty intersection. Then for any point $P \in S_1 \cap S_2$, there is a characteristic curve of the vector field $\mathcal{F}_{a,b,c}$, going through P, whose range is contained in $S_1 \cap S_2$.

Proof. (i) Let a point $P = \begin{pmatrix} x_0 \\ y_0 \\ z_0 \end{pmatrix}$, where $z_0 = u(x_0, y_0)$, be a point of the graph S of u;

in particular, $\begin{pmatrix} x_0 \\ y_0 \end{pmatrix} \in \Omega$. By Proposition 5.2.1, there is a characteristic curve

$$\gamma(t) = \begin{pmatrix} \bar{x}(t) \\ \bar{y}(t) \\ \bar{z}(t) \end{pmatrix}, \quad (t \in K),$$

on an open interval K about $t_0 \in \mathbf{R}$ which passes through P at $t = t_0$.

Since the point $\begin{pmatrix} \tilde{x}(t_0) \\ \tilde{y}(t_0) \end{pmatrix} = \begin{pmatrix} x_0 \\ y_0 \end{pmatrix}$ belongs to Ω, we have, because the domain Ω is open and the curve y is continuous, that there is an open interval $J \subseteq K$ about t_0 such that

$$\begin{pmatrix} \tilde{x}(t) \\ \tilde{y}(t) \end{pmatrix} \in \Omega$$

for all $t \in J$, or, for short,

$$\begin{pmatrix} \tilde{x} \\ \tilde{y} \end{pmatrix} \in \Omega$$

on J. In effect, as u is a solution of the PDE in question,

$$a(\tilde{x}, \tilde{y}, u(\tilde{x}, \tilde{y}))u_x(\tilde{x}, \tilde{y}) + b(\tilde{x}, \tilde{y}, u(\tilde{x}, \tilde{y}))u_y(\tilde{x}, \tilde{y}) \tag{7.2.2}$$
$$= c(\tilde{x}, \tilde{y}, u(\tilde{x}, \tilde{y}))$$

identically on J.

Consider the function

$$U(t) = \tilde{z}(t) - u(\tilde{x}(t), \tilde{y}(t)) \tag{7.2.3}$$

on J. Observe that

$$U(t_0) = \tilde{z}(t_0) - u(\tilde{x}(t_0), \tilde{y}(t_0))$$
$$= u(x_0, y_0) - u(x_0, y_0) = 0.$$

We are going to construct

a certain explicit first-order *ODE* which is satisfied by the function U on J and by (surprise, surprise) the zero function on J, which also takes the zero value at $t = t_0$.

We then apply the uniqueness part of the Peano–Pickard–Lindelöf theorem to use the fact that U and the zero function must coincide on an open subinterval of J about t_0.

So let us differentiate both parts of (7.2.3) by t, thereby getting (we shall again switch to shorter symbols) that

$$U' = \tilde{z}' - u_x(\tilde{x}, \tilde{y})\tilde{x}' - u_y(\tilde{x}, \tilde{y})\tilde{y}',$$

whence, since y is characteristic curve,

$$U' = c(\tilde{x}, \tilde{y}, \tilde{z}) - u_x(\tilde{x}, \tilde{y})a(\tilde{x}, \tilde{y}, \tilde{z}) - u_y(\tilde{x}, \tilde{y})b(\tilde{x}, \tilde{y}, \tilde{z}).$$

Now replace $\tilde{z}$ in the last equation with $U + u(\tilde{x}, \tilde{y})$:

$$U' = c(\tilde{x}, \tilde{y}, U + u(\tilde{x}, \tilde{y}))$$
$$- u_x(\tilde{x}, \tilde{y})a(\tilde{x}, \tilde{y}, U + u(\tilde{x}, \tilde{y}))$$
$$- u_y(\tilde{x}, \tilde{y})b(\tilde{x}, \tilde{y}, U + u(\tilde{x}, \tilde{y})).$$

So U is a solution of the ODE

$$W' = c(\tilde{x}, \tilde{y}, W + u(\tilde{x}, \tilde{y})) \qquad (7.2.4)$$
$$- u_x(\tilde{x}, \tilde{y})a(\tilde{x}, \tilde{y}, W + u(\tilde{x}, \tilde{y}))$$
$$- u_y(\tilde{x}, \tilde{y})b(\tilde{x}, \tilde{y}, W + u(\tilde{x}, \tilde{y})),$$

for an unknown function $W = W(t)$. But, by (7.2.2), the zero function on J is also a solution of this ODE, as we observed above.

Rewrite the ODE (7.2.4) in the form

$$W' = F(t, W).$$

By the conditions, there is $\varepsilon > 0$ such that the functions

$$a, b, c, a_z, b_z, c_z$$

exist and are continuous on the open ball $B_\varepsilon(P)$ of radius ε centered at P. Consider the smooth curve

$$\omega(t) = \begin{pmatrix} \tilde{x}(t) \\ \tilde{y}(t) \\ u(\tilde{x}(t), \tilde{y}(t)) \end{pmatrix}, \quad (t \in J)$$

on J. By continuity of the component functions of the curve ω, there exists $\delta > 0$ such that whenever $|t - t_0| < \delta$, then

$$t \in J, \ \begin{pmatrix} \tilde{x}(t) \\ \tilde{y}(t) \end{pmatrix} \in \Omega, \quad \text{and} \quad \omega(t) \in B_{\varepsilon/2}(P).$$

Accordingly,

$$B_{\varepsilon/2}(\omega(t)) \subseteq B_\varepsilon(P)$$

for all $t \in (t_0 - \delta, t_0 + \delta)$. This implies that

$$\begin{pmatrix} \tilde{x}(t) \\ \tilde{y}(t) \end{pmatrix} \in \Omega \quad \text{and} \quad \begin{pmatrix} \tilde{x}(t) \\ \tilde{y}(t) \\ W + u(\tilde{x}(t), \tilde{y}(t)) \end{pmatrix} \in B_\varepsilon(P)$$

for all $\begin{pmatrix} t \\ W \end{pmatrix}$ in the open neighborhood

$$\Pi = (t_0 - \delta, t_0 + \delta) \times (-\varepsilon/2, \varepsilon/2)$$

of the point $\begin{pmatrix} t_0 \\ W_0 \end{pmatrix} = \begin{pmatrix} t_0 \\ 0 \end{pmatrix}$. It is then easy to see that both the function $F(t, W)$ and its partial derivative $F_W(t, W)$ by W exist and are continuous on Π.

By the uniqueness part of the Peano–Pickard–Lindelöf theorem applied to the function $F = F(t, W)$ on Π and the ODE

$$W' = F(t, W),$$

we then obtain that the function $U = U(t)$ coincides with the zero function on the open interval $I = (t_0 - \delta, t_0 + \delta) \subseteq J$ about t_0. Therefore,

$$U(t) = 0$$

for all $t \in I$, or

$$\tilde{z}(t) = u\big(\tilde{x}(t), \tilde{y}(t)\big)$$

for all $t \in I$. This means that the curve

$$\eta(t) = \begin{pmatrix} \tilde{x}(t) \\ \tilde{y}(t) \\ \tilde{z}(t) \end{pmatrix}, \quad (t \in I)$$

lies entirely on S, and S is indeed a union of characteristic curves of our PDE, as claimed.

(ii) Let $\begin{pmatrix} x_0 \\ y_0 \end{pmatrix}$ be any point of Ω. Clearly, the proof of (i) implies that there is a solution of the system

$$\begin{cases} x'(t) = a(\, x(t),\, y(t),\, u(x(t), y(t))\,), \\ y'(t) = b(\, x(t),\, y(t),\, u(x(t), y(t))\,) \end{cases} \tag{7.2.5}$$

on a suitable open interval about t_0, subject to the initial conditions

$$x(t_0) = x_0 \quad \text{and} \quad y(t_0) = y_0.$$

Conversely, suppose that a pair of functions $x(t), y(t)$ gives a solution to the system (7.2.5) on an open interval I about t_0, meets the initial conditions

$$x(t_0) = x_0 \quad \text{and} \quad y(t_0) = y_0,$$

and the range of the corresponding map

$$t \mapsto \begin{pmatrix} x(t) \\ y(t) \end{pmatrix}, \quad (t \in I)$$

is contained in Ω. We claim that the curve

$$\gamma(t) = \begin{pmatrix} x(t) \\ y(t) \\ u(x(t), y(t)) \end{pmatrix}, \quad (t \in I)$$

is a characteristic curve of the vector field $\mathcal{F}_{a,b,c}$, which lies entirely in the graph of u. We have then to show that

$$\gamma'(t) = \mathcal{F}_{a,b,c}(\gamma(t)),$$

for all $t \in I$. Since, by (7.2.5),

$$\begin{cases} x'(t) = a(\gamma(t)), \\ y'(t) = b(\gamma(t)), \end{cases} \quad (t \in I),$$

it remains to show that

$$[u(x(t), y(t))]' = c(\gamma(t))$$
$$= c(x(t), y(t), u(x(t), y(t)))$$

on I. It is easy: indeed, for all $t \in I$,

$$[u(x(t), y(t))]' = u_x(x(t), y(t))x'(t) + u_y(x(t), y(t))y'(t)$$
$$= a(\gamma(t))u_x(x(t), y(t)) + b(\gamma(t))u_y(x(t), y(t))$$
$$= c(\gamma(t)),$$

where the last equality is justified by the fact that u is a solution of the PDE under consideration on Ω.

(iii) By (i), there is a characteristic curve

$$\gamma_k : (-\delta_k, \delta_k) \to S_k = \text{Graph}(u_k),$$

where $\delta_k > 0$, of the vector field $\mathcal{F}_{a,b,c}$ such that $\gamma_k(0) = P$ $(k = 1, 2)$. Then, by part (i) of Proposition 5.2.1,

$$\gamma_1 \equiv \gamma_2$$

on the open interval $(-\delta, \delta)$, where $\delta = \min(\delta_1, \delta_2)$. So one of the curves γ_1, γ_2 is a desired characteristic curve which lies entirely in $S_1 \cap S_2$. $\qquad\square$

Recall that the maximum (max) norm $\| \cdot \|_\infty$ on the arithmetic space $\mathbf{R}^n$ is defined by

$$\left\| \begin{pmatrix} x_1 \\ x_2 \\ \vdots \\ x_n \end{pmatrix} \right\|_\infty = \max(|x_1|, |x_2|, \ldots, |x_n|), \qquad \begin{pmatrix} x_1 \\ x_2 \\ \vdots \\ x_n \end{pmatrix} \in \mathbf{R}^n.$$

To remind the reader, if $N = \| \cdot \| : \mathbf{R}^n \to \mathbf{R}$ is any norm on $\mathbf{R}^n$, then the set

$$B_N(P, r) = \{Q \in \mathbf{R}^n : \|P - Q\| < r\}$$

where $P \in \mathbf{R}^n$ and $r > 0$, is called on an open ball of radius r centered at P with regard to the norm N. Due to equivalence of norms on $\mathbf{R}^n$ [9, pp. 106–107], a subset O of $\mathbf{R}^n$ is open if and only if for every point $P \in O$ there is $\delta > 0$ such that $B_N(P, \delta) \subseteq O$.

Theorem 7.2.2 (Method of characteristics for first-order quasilinear equations). *Let Λ be an open set in $\mathbf{R}^3$, let functions $a, b, c \in C^1(\Lambda)$ be continuously differentiable everywhere on Λ, and let f, g, h be continuously differentiable functions on an open interval I in $\mathbf{R}$ such that the curve*

$$\Gamma(s) = \begin{pmatrix} f(s) \\ g(s) \\ h(s) \end{pmatrix}, \qquad (s \in I)$$

lies entirely in Λ. Write $\tilde{\Gamma}$ for the orthogonal projection of Γ on the xy-plane.

For every $s \in I$, let J_s be an open interval about 0 and let functions $x_s(t), y_s(t), z_s(t) : J_s \to \mathbf{R}$ in variable t give a solution to the characteristic system

$$\begin{cases} x'(t) = a(x(t), y(t), z(t)), \\ y'(t) = b(x(t), y(t), z(t)), \\ z'(t) = c(x(t), y(t), z(t)) \end{cases}$$

of the vector field $\mathcal{F}_{a,b,c}$ on Λ with the initial conditions

$$x(0) = f(s), \quad y(0) = g(s), \quad z(0) = h(s).$$

Suppose that the intervals J_s are chosen so that the set

$$\Theta = \bigcup_{s \in I} \{s\} \times J_s$$

is a domain in $\mathbf{R}^2$. Set then

$$X(s, t) = x_s(t), \quad Y(s, t) = y_s(t), \quad Z(s, t) = z_s(t)$$

for all $\begin{pmatrix} s \\ t \end{pmatrix} \in \Theta$. Suppose further that the functions $X(s,t), Y(s,t), Z(s,t)$ are continuously differentiable on Θ and that the function

$$\Psi\begin{pmatrix} s \\ t \end{pmatrix} = \begin{pmatrix} X(s,t) \\ Y(s,t) \end{pmatrix}, \quad \begin{pmatrix} s \\ t \end{pmatrix} \in \Theta$$

on Θ is C^1-invertible, that is, there are continuously differentiable functions $S(x,y)$ and $T(x,y)$ on $\Omega = \Psi(\Theta)$ such that

$$\Psi^{-1}\begin{pmatrix} x \\ y \end{pmatrix} = \begin{pmatrix} S(x,y) \\ T(x,y) \end{pmatrix}, \quad \begin{pmatrix} x \\ y \end{pmatrix} \in \Omega.$$

Then:

(i) *The function*

$$v(x,y) = Z(S(x,y), T(x,y))$$

is a solution of the Cauchy problem for the PDE

$$a(x,y,u)u_x + b(x,y,u)u_y = c(x,y,u) \tag{7.2.6a}$$

on Ω with the initial condition

$$u(f(s),g(s)) = h(s), \tag{7.2.6b}$$

where s runs over I (existence);

(ii) *Moreover, if a function $w \in C^1(\Omega)$ is also a solution of the PDE (7.2.6a) on Ω and satisfies the initial condition (7.2.6b), then*

$$v \equiv w$$

on a suitable subdomain Ω_w of Ω such that Ω_w contains $\widetilde{\Gamma}$; accordingly, the Cauchy problem (7.2.6) has a unique solution up to equivalence (uniqueness).

Proof. First, note that due to $\begin{pmatrix} X(s,t) \\ Y(s,t) \\ Z(s,t) \end{pmatrix} \in \Lambda$ for all $\begin{pmatrix} s \\ t \end{pmatrix} \in \Theta$, we get that for every

$\begin{pmatrix} x \\ y \end{pmatrix} \in \Omega$, the point

$$\begin{pmatrix} x \\ y \\ Z(S(x,y),T(x,y)) \end{pmatrix} = \begin{pmatrix} X(S(x,y),T(x,y)) \\ Y(S(x,y),T(x,y)) \\ Z(S(x,y),T(x,y)) \end{pmatrix} \in \Lambda$$

is in Λ, and, in effect,

$$\Omega \subseteq \mathrm{proj}_{x,y}(\Lambda).$$

Another must-check relation which follows from the conditions is

$$\tilde{\Gamma} \subseteq \Omega,$$

for indeed

$$\Omega \ni \Psi \begin{pmatrix} s \\ 0 \end{pmatrix} = \begin{pmatrix} X(s,0) \\ Y(s,0) \end{pmatrix} = \begin{pmatrix} f(s) \\ g(s) \end{pmatrix}$$

for all $s \in I$.

Next, since, by the construction, the Monge surface

$$\begin{cases} x = X(s,t), \\ y = Y(s,t), \quad \begin{pmatrix} s \\ t \end{pmatrix} \in \Theta \\ z = Z(s,t), \end{cases}$$

is, again due to invertibility of the function

$$\Psi \begin{pmatrix} s \\ t \end{pmatrix} = \begin{pmatrix} X(s,t) \\ Y(s,t) \end{pmatrix}, \quad \begin{pmatrix} s \\ t \end{pmatrix} \in \Omega,$$

the graph of the C^1-function

$$v(x,y) = Z(S(x,y), T(x,y))$$

on Ω (Proposition 7.1.2), and this graph is "made" up of characteristic curves of the vector field $\mathcal{F}_{a,b,c}$, we conclude that the function v is a solution of the PDE (7.2.6a) on Ω (Proposition 7.1.1). Also, by the construction

$$u(X(s,t), Y(s,t)) = Z(s,t)$$

everywhere on Θ, and hence

$$u(X(s,0), Y(s,0)) = Z(s,0),$$

or

$$u(f(s), g(s)) = h(s)$$

for all $s \in I$. This proves (i).

Consider an *arbitrary* solution $u \in C^1(\Omega)$ of the PDE (7.2.6a) on Ω satisfying Graph$(u) \subseteq \Lambda$. Before moving on to the proof of (ii), let us address an important question concerning the graph Graph(u) of u.

As we know (Proposition 7.2.1), the graph of u is "made" up of characteristic curves of the vector field $\mathcal{F}_{a,b,c}$. Let M, N be points on the graph of u that are close enough to each other. Now a reasonable question is: can then we prove that there are characteristic curves γ_M, γ_N of our vector field, both passing through the corresponding points at $t = 0$,

$$\gamma_M(0) = M \quad \text{and} \quad \gamma_N(0) = N,$$

both lying entirely in the graph of u, and both having the *same* domain $(-\delta, \delta)$, where $\delta > 0$? As the reader may have already guessed, once this question is answered, we will be able to prove (ii) quickly enough.

We use the theorem on continuous dependence on initial values (see p. 89). Take a point $\begin{pmatrix} \xi \\ \eta \end{pmatrix} \in \Omega$, a real number $\tau \in \mathbf{R}$, and consider the initial value problem

$$\begin{cases} x'(t) = a(x(t), y(t), u(x(t), y(t))), \\ y'(t) = b(x(t), y(t), u(x(t), y(t))), \\ x(\tau) = \xi, \\ y(\tau) = \eta, \end{cases} \tag{7.2.7a}$$

where we also require that

$$\beta(t) = \begin{pmatrix} x(t) \\ y(t) \end{pmatrix} \in \Omega \tag{7.2.7b}$$

for all points t in the domain of β (as usual, an open interval in $\mathbf{R}$ about τ; cf. the formulation of part (ii) of Proposition 7.2.1). The requirement (7.2.7b) is plainly necessary since we are going to apply Proposition 7.2.1 and, moreover, it enables us to conform with the conditions of the Peano–Pickard–Lindelöf theorem: indeed, we can think of the system (7.2.7a) as a system of the form

$$\begin{cases} x'(t) = A(t, x(t), y(t)), \\ y'(t) = B(t, x(t), y(t)), \end{cases}$$

where the functions A, B are defined on the domain $\mathbf{R} \times \Omega$ in $\mathbf{R}^3$.

Now, by Proposition 7.2.1, any solution of the initial value problem (7.2.7) on an open interval about τ is a characteristic curve of the vector field $\mathcal{F}_{a,b,c}(\Lambda)$, whose range is contained in the graph of u, and which passes through the point $\begin{pmatrix} \xi \\ \eta \\ u(\xi, \eta) \end{pmatrix}$ on the graph of u. Let then

$$\varphi_{\tau, \xi, \eta}(t) = \begin{pmatrix} x(t) \\ y(t) \end{pmatrix}$$

be a nonextendable solution of the initial value problem (7.2.7) on a certain open interval $I_{\xi,\eta}$ in **R** about τ. This defines a function

$$\varphi(t, \tau, \xi, \eta) = \varphi_{\tau,\xi,\eta}(t)$$

on $\mathbf{R} \times (\mathbf{R} \times \Omega)$, *whose domain is,* by the theorem on continuous dependence on initial values, *an open set in* $\mathbf{R}^4$.

Pick up a point $Q = \begin{pmatrix} x_0 \\ y_0 \end{pmatrix} \in \Omega$ and consider the point $\begin{pmatrix} 0 \\ 0 \\ x_0 \\ y_0 \end{pmatrix}$ of $\mathbf{R}^4$ from the

domain of φ. As the domain of the function φ is open, there is $\delta > 0$ such that the open set

$$U_\delta = \left\{ \begin{pmatrix} t \\ \tau \\ x \\ y \end{pmatrix} \in \mathbf{R}^4 : \max \begin{pmatrix} |t - 0| \\ |\tau - 0| \\ |x - x_0| \\ |y - y_0| \end{pmatrix} < \delta \right\}$$

is contained in the domain of φ; here,

$$\max \begin{pmatrix} |t - 0| \\ |\tau - 0| \\ |x - x_0| \\ |y - y_0| \end{pmatrix}$$

represents distance between the points

$$\begin{pmatrix} t \\ \tau \\ x \\ y \end{pmatrix} \quad \text{and} \quad \begin{pmatrix} 0 \\ 0 \\ x_0 \\ y_0 \end{pmatrix}$$

with regard to the max norm on $\mathbf{R}^4$ we have discussed above. By making δ smaller if necessary, we can assume that the open set

$$W_{Q,u} = \left\{ \begin{pmatrix} x \\ y \end{pmatrix} \in \mathbf{R}^2 : \max \begin{pmatrix} |x - x_0| \\ |y - y_0| \end{pmatrix} < \delta \right\}$$

is contained in Ω.

To complete the argument: whenever $\begin{pmatrix} \xi \\ \eta \end{pmatrix} \in W_{Q,u}$, then $\begin{pmatrix} t \\ 0 \\ \xi \\ \eta \end{pmatrix} \in U_\delta$ for all t with

$|t| < \delta$, which means that the point $\begin{pmatrix} t \\ 0 \\ \xi \\ \eta \end{pmatrix}$ belongs to the domain of φ, and which, in turn, means that the value

$$\varphi_{0,\xi,\eta}(t)$$

exists.

In other words, for every point $\begin{pmatrix} \xi \\ \eta \end{pmatrix}$ of the open neighborhood $W_{Q,u}$ of Q, there is a characteristic curve

$$t \mapsto \begin{pmatrix} x(t) \\ y(t) \\ u(x(t), y(t)) \end{pmatrix} \tag{7.2.8}$$

whose range is contained in the graph of u, which passes through the point $\begin{pmatrix} \xi \\ \eta \\ u(\xi, \eta) \end{pmatrix}$ at $t = 0$, and whose domain is equal to the open interval $(-\delta, \delta)$, as aspired.

Now we are ready to prove part (ii) of the theorem. Let $w \in C^1(\Omega)$ be another solution of the Cauchy problem from part (ii), which satisfies the PDE (7.2.6a) on Ω and the initial condition (7.2.6b). In particular, for every $s \in I$, the point $\Gamma(s)$ belongs both to the graph of u and the graph of w.

Let σ be a point of I and let $Q = \begin{pmatrix} f(\sigma) \\ g(\sigma) \end{pmatrix}$. Due to continuity of the functions f, g, there is a positive number $\alpha > 0$ such that

$$(\sigma - \alpha, \sigma + \alpha) \subseteq I,$$

and

$$\tilde{\Gamma}(s) \in W_{Q,v} \cap W_{Q,w},$$

for all $s \in (\sigma - \alpha, \sigma + \alpha)$. By the construction of the open sets $W_{Q,v}$ and $W_{Q,w}$, there are $\delta_1, \delta_2 > 0$ such that for every $s \in (\sigma - \alpha, \sigma + \alpha)$,

... there is a characteristic curve $\gamma_{1,s}$ (resp., $\gamma_{2,s}$) having the form (7.2.8) for $u = v$ (resp., $u = w$),

... whose domain is equal to $(-\delta_1, \delta_1)$ (resp., to $(-\delta_2, \delta_2)$),

... which passes through the point $\Gamma(s)$ at $t = 0$,

... and which lies entirely in the graph of v (resp., of w).

As $\gamma_{1,s}(0) = \gamma_{2,s}(0)$, then, by part (i) of Proposition 5.2.1, we get that

$$\gamma_{1,s} \equiv \gamma_{2,s}$$

identically on $(-\delta, \delta)$ for all $s \in (\sigma - \alpha, \sigma + \alpha)$, where

$$\delta = \min(\delta_1, \delta_2).$$

Write R_σ for the open rectangle

$$(\sigma - \alpha, \sigma + \alpha) \times (-\delta, \delta),$$

and set

$$\Theta_\sigma = \Theta \cap R_\sigma;$$

clearly, Θ_σ is a domain in the st-plane. We then have that

$$\Theta_\sigma = \left(\bigcup_{s \in I} \{s\} \times J_s \right) \cap \left(\bigcup_{s \in (\sigma - \alpha, \sigma + \alpha)} \{s\} \times (-\delta, \delta) \right)$$

$$= \bigcup_{s \in (\sigma - \alpha, \sigma + \alpha)} \{s\} \times (J_s \cap (-\delta, \delta)).$$

Let $s \in (\sigma - \alpha, \sigma + \alpha)$. Then due to

$$\gamma_{1,s}(0) = \gamma_{2,s}(0) = \begin{pmatrix} x_s(0) \\ y_s(0) \\ z_s(0) \end{pmatrix} \quad (= \Gamma(s)),$$

we, again by Proposition 5.2.1 (i), obtain that

$$\gamma_{1,s}(t) = \gamma_{2,s}(t) = \begin{pmatrix} x_s(t) \\ y_s(t) \\ z_s(t) \end{pmatrix} = \begin{pmatrix} X(s,t) \\ Y(s,t) \\ Z(s,t) \end{pmatrix}$$

for all $t \in J_s \cap (-\delta, \delta)$, whence, after recalling the construction of the curves $\gamma_{1,s}$ and $\gamma_{2,s}$,

$$Z(s,t) = \underbrace{v(X(s,t), Y(s,t)) = w(X(s,t), Y(s,t))}$$

for *all* $\begin{pmatrix} s \\ t \end{pmatrix} \in \Theta_\sigma$.

We therefore conclude that the functions v and w coincide,

$$v \equiv w,$$

on the domain $\Omega_\sigma = \Psi(\Theta_\sigma)$ in $\mathbf{R}^2$, an open neighborhood of the point $\begin{pmatrix} f(\sigma) \\ g(\sigma) \end{pmatrix}$ on the curve $\tilde{\Gamma}$.

Finally, we have that the functions v and w coincide,

$$u \equiv w,$$

on the open set

$$\Omega_w = \bigcup_{\sigma \in I} \Omega_\sigma,$$

which is also connected, since whenever σ moves over I in a continuous fashion, the corresponding connected set Ω_σ moves along the continuous curve $\tilde{\Gamma}$. The theorem is proven. $\qquad \square$

Remark 7.2.3. We shall keep the notation and assumptions of Theorem 7.2.2.

(a) First, observe that if the function Ψ is C^1-invertible, its Jacobian determinant,

$$\det J_\Psi(s,t) = \frac{\partial(X,Y)}{\partial(s,t)} = \begin{vmatrix} X_s & Y_s \\ X_t & Y_t \end{vmatrix} \neq 0,$$

must be nonzero everywhere on Θ. As, by the conditions of the theorem, the functions X, Y are continuously differentiable on Θ and as, for instance,

$$X(s,0) = f(s) \quad \text{and} \quad X(s,t) = x_s(t)$$

for all $\begin{pmatrix} s \\ t \end{pmatrix} \in \Theta$, we get that

$$X_s(s,0) = f'(s)$$

and

$$X_t(s,0) = x'_s(0) = a(x_s(0), y_s(0), z_s(0))$$
$$= a(f(s), g(s), h(s))$$

for all $s \in I$. Similarly,

$$Y_s(s,0) = g'(s)$$

and

$$Y_t(s,0) = y'_s(0) = b(f(s), g(s), h(s)),$$

for all $s \in I$. It follows that

$$\det J_\Psi(s,0) = \begin{vmatrix} X_s(s,0) & Y_s(s,0) \\ X_t(s,0) & Y_t(s,0) \end{vmatrix}$$

$$= \begin{vmatrix} f'(s) & g'(s) \\ a(f(s), g(s), h(s)) & b(f(s), g(s), h(s)) \end{vmatrix}$$

$$= \begin{vmatrix} f'(s) & g'(s) \\ a(\Gamma(s)) & b(\Gamma(s)) \end{vmatrix} \neq 0$$

for all $s \in I$. Recall that we have already seen this condition, which is called the transversality condition, as one of the hypotheses in Theorem 6.4.3; however, this time the condition appears on the scene in a less forceful manner.

So we can state that the transversality condition holds for all points of I. Extending a remark on the transversality condition from Chapter 6, we see that

$$\mathrm{rank}\begin{pmatrix} f'(s) & g'(s) & h'(s) \\ a(\Gamma(s)) & b(\Gamma(s)) & c(\Gamma(s)) \end{pmatrix} = 2 \qquad (7.2.9)$$

for all $s \in I$, or, in other words, the tangent vector $\Gamma'(s)$ to Γ at s and the field vector $\mathcal{F}_{a,b,c}(\Gamma(s))$ at $\Gamma(s)$ are never parallel,

$$\Gamma'(s) = \begin{pmatrix} f'(s) \\ g'(s) \\ h'(s) \end{pmatrix} \nparallel \begin{pmatrix} a(\Gamma(s)) \\ b(\Gamma(s)) \\ c(\Gamma(s)) \end{pmatrix} = \mathcal{F}_{a,b,c}(\Gamma(s)),$$

as s runs over I. In still other words,

the tangent vector $\Gamma'(s)$ never parallel to the characteristic direction at $\Gamma(s)$,

for all $s \in I$, while, quite contrary, along any characteristic curve the tangent vector always equals the characteristic direction at any point of its range.

To sum up, the validity of the transversality condition

$$\begin{vmatrix} f'(s) & g'(s) \\ a(f(s), g(s), h(s)) & b(f(s), g(s), h(s)) \end{vmatrix} \neq 0 \qquad (7.2.10)$$

for all $s \in I$ is *necessary* for the method of characteristics to work.

(b) A smooth curve $\Gamma(s) = \begin{pmatrix} f(s) \\ g(s) \\ h(s) \end{pmatrix}$ which satisfies the condition (7.2.9) at each point of its domain can be, in view of our discussion in (a), called be *anticharacteristic* (for the lack of a better/commonly used term). $\triangle$

Definition (Equivalence of local solutions of a Cauchy problem for a first-order PDE). Let

$$\begin{cases} \mathcal{E}(u) = 0, \\ u(f(s), g(s)) = h(s), \quad (s \in I), \end{cases} \qquad (7.2.11)$$

be a Cauchy problem for a first-order partial differential equation $\mathcal{E}(u) = 0$ on an open set $\mathcal{O}$ of $\mathbf{R}^2$ and let s_0 be an inner point of the interval I. Write Γ for the initial curve on I with the component functions f, g, h.

We say that local solutions $u_1 \in C^1(\Omega_1)$ and $u_2 \in C^1(\Omega_2)$ of the Cauchy problem (7.2.11) at $s = s_0$ are (locally) *equivalent*, symbolically

$$u_1 \sim_{\mathrm{loc}} u_2,$$

if there is a positive number $\alpha > 0$ such that the interval

$$J_\alpha = (s_0 - \alpha, s_0 + \alpha)$$

is contained in I and u_1, u_2 are equivalent solutions of the Cauchy problem (7.2.11) restricted to the interval J_α, that is, there exists a domain Π of $\mathbf{R}^2$ such that

$$\tilde{\Gamma}(J_\alpha) \subseteq \Pi \subseteq \Omega_1 \cap \Omega_2,$$

where $\tilde{\Gamma}$ is the orthogonal projection of Γ on the xy-plane, and

$$u_1(x, y) = u_2(x, y)$$

for all $\begin{pmatrix} x \\ y \end{pmatrix} \in \Pi$. $\triangle$

Claim 7.2.4. *The relation $\sim_{loc}$ is an equivalence relation on the set of local solutions of the Cauchy problem (7.2.11) at $s = s_0$.*

The proof is quite easy, and is left to the reader.

Based on our discussion in part (a) of Remark 7.2.3, we obtain the following, local version of Theorem 7.2.2.

Proposition 7.2.5 (Method of characteristics: local version). *Let Λ be an open set in $\mathbf{R}^3$, let functions $a, b, c \in C^1(\Lambda)$ be continuously differentiable everywhere on Λ, and let f, g, h be continuously differentiable functions on an open interval I in $\mathbf{R}$ such that the curve*

$$\Gamma(s) = \begin{pmatrix} f(s) \\ g(s) \\ h(s) \end{pmatrix}, \quad (s \in I)$$

lies entirely in Λ.

Suppose that the transversality condition

$$\begin{vmatrix} f'(s_0) & g'(s_0) \\ a(f(s_0), g(s_0), h(s_0)) & b(f(s_0), g(s_0), h(s_0)) \end{vmatrix} \neq 0$$

holds for a certain point $s_0 \in I$. Then the Cauchy problem

$$\begin{cases} a(x, y, u)u_x + b(x, y, u)u_y = c(x, y, u), \\ u(f(s), g(s)) = h(s), \quad (s \in I) \end{cases}$$

has a local solution at $s = s_0$, which is unique up to equivalence of local solutions.

A nice feature of the local version of the method of characteristics as presented in Proposition 7.2.5 is that the verification of the corresponding conditions needs considerably less effort in comparison with all similar results we have obtained above (Proposition 6.4.1, Theorem 6.4.3, and Theorem 7.2.2). For instance, application of the main theorem of the method may meet, to continue to follow the notation of Theorem 7.2.2, difficulties in the verification that the map Ψ is C^1-invertible, in choosing intervals J_s in an appropriate way to form a domain, and so on.

Proof of Proposition 7.2.5. Take a point $\begin{pmatrix} \xi \\ \eta \\ \sigma \end{pmatrix}$ of Λ and consider the characteristic system

$$\begin{cases} x'(t) = a(x(t), y(t), z(t)), \\ y'(t) = b(x(t), y(t), z(t)), \\ z'(t) = c(x(t), y(t), z(t)), \end{cases} \qquad (7.2.12\text{a})$$

subject to the initial conditions

$$x(0) = \xi, \quad y(0) = \eta, \quad z(0) = \sigma, \qquad (7.2.12\text{b})$$

assuming, in the context of Peano–Pickard–Lindelöf theorem, that the functions in the right-hand sides of the equations of the system are defined on the open set $\mathbf{R} \times \Lambda$ in $\mathbf{R}^4$.

Using the theorem on differentiability with respect to initial values (for $\tau_0 = 0$; see p. 89), consider the function

$$\boldsymbol{\psi}(t, \xi, \eta, \sigma) = \begin{pmatrix} \psi_1(t, \xi, \eta, \sigma) \\ \psi_2(t, \xi, \eta, \sigma) \\ \psi_3(t, \xi, \eta, \sigma) \end{pmatrix}$$

on $\mathbf{R} \times \Lambda$ associated with (7.2.12), where ψ_i are real-valued functions, whose definition is given in the formulation of the theorem.

Now the point $\begin{pmatrix} 0 \\ f(s_0) \\ g(s_0) \\ h(s_0) \end{pmatrix}$ is in the domain of $\boldsymbol{\psi}$, which is, by the theorem, an open set in $\mathbf{R}^4$. Accordingly, there is a positive number $\delta > 0$ such that the open set

$$V_\delta = \left\{ \begin{pmatrix} t \\ x \\ y \\ z \end{pmatrix} \in \mathbf{R}^4 : \max \begin{pmatrix} |t - 0| \\ |x - f(s_0)| \\ |y - g(s_0)| \\ |z - h(s_0)| \end{pmatrix} < \delta \right\}$$

is contained in dom($\boldsymbol{\psi}$). By continuity of f, g, h, there exists $\alpha > 0$ such that

$$|s - s_0| < \alpha \Rightarrow \max \begin{pmatrix} |f(s) - f(s_0)| \\ |g(s) - g(s_0)| \\ |h(s) - h(s_0)| \end{pmatrix} < \delta$$

for all $s \in I$. Set

$$Y = (s_0 - \alpha, s_0 + \alpha) \times (-\delta, \delta).$$

Consequently, whenever $\begin{pmatrix} s \\ t \end{pmatrix} \in Y$, then the value

$$\boldsymbol{\psi}(t, f(s), g(s), h(s))$$

exists. Set then further

$$X(s, t) = \psi_1(t, f(s), g(s), h(s)),$$
$$Y(s, t) = \psi_2(t, f(s), g(s), h(s)),$$
$$Z(s, t) = \psi_3(t, f(s), g(s), h(s)),$$

where $\begin{pmatrix} s \\ t \end{pmatrix}$ changes over Y. By applying the theorem on differentiability with respect to initial values again, we see that given any s in $(s_0 - \alpha, s_0 + \alpha)$, the map

$$t \mapsto \begin{pmatrix} X(s, t) \\ Y(s, t) \\ Z(s, t) \end{pmatrix},$$

is a characteristic curve of the vector field $\mathcal{F}_{a,b,c}$ which passes through the point $\Gamma(s)$ at $t = 0$, and that the functions $X(s, t), Y(s, t), Z(s, t)$ are continuously differentiable on Y, since so are the functions ψ_i on $\mathrm{dom}(\boldsymbol{\psi})$ and the functions f, g, h on I.

Running along the lines of Theorem 7.2.2, define the map

$$\Psi \begin{pmatrix} s \\ t \end{pmatrix} = \begin{pmatrix} X(s, t) \\ Y(s, t) \end{pmatrix}, \quad \begin{pmatrix} s \\ t \end{pmatrix} \in Y$$

on Y. As is discussed in part (a) of Remark 7.2.3, we have that

$$J_\Psi(s_0, 0) = \begin{vmatrix} f'(s_0) & g'(s_0) \\ a(f(s_0), g(s_0), h(s_0)) & b(f(s_0), g(s_0), h(s_0)) \end{vmatrix}.$$

Hence, by the hypothesis,

$$\det J_\Psi(s_0, 0) \neq 0.$$

Accordingly, by the Inverse Function Theorem, the map Ψ is locally C^1-invertible near the point $\begin{pmatrix} s_0 \\ 0 \end{pmatrix}$ of the st-plane, say, its restriction on an open rectangle

$$Y_1 = (s_0 - \alpha_1, s_0 + \alpha_1) \times (-\delta_1, \delta_1),$$

where $\alpha_1, \delta_1 > 0$, which is contained in Y, is C^1-invertible. The result then, after applying Theorem 7.2.2 to the restriction of Γ on $(s_0 - \alpha_1, s_0 + \alpha_1)$, to the domain Y_1 in the st-plane and to the C^1-invertible map $\Psi|_{Y_1}$, follows easily. $\qquad\square$

The remaining case in our study of Cauchy problems for first-order quasilinear PDEs is the case when the transversality condition fails for all points of the domain I of the initial curve Γ:

$$\begin{vmatrix} f'(s) & g'(s) \\ a(f(s), g(s), h(s)) & b(f(s), g(s), h(s)) \end{vmatrix} = 0 \qquad (7.2.13)$$

for *all* $s \in I$. As was remarked above, this is most notably the case for any characteristic curve $\Gamma(s)$ of the vector field $\mathcal{F}_{a,b,c}$, for which we have that

$$\Gamma'(s) = \mathcal{F}_{a,b,c}(\Gamma(s)) \iff$$
$$\begin{pmatrix} f'(s) \\ g'(s) \\ h'(s) \end{pmatrix} = \begin{pmatrix} a(f(s), g(s), h(s)) \\ b(f(s), g(s), h(s)) \\ c(f(s), g(s), h(s)) \end{pmatrix},$$

for all $s \in I$, and in this case the determinant in (7.2.13) will simply have two equal rows.

Further examples are given by curves $\Gamma(s)$ whose tangent vectors $\Gamma'(s)$ are parallel to the field vector $\mathcal{F}_{a,b,c}(\Gamma(s))$ at any point of their ranges. For any such curve, we shall have that

$$\operatorname{rank}\begin{pmatrix} f'(s) & g'(s) & h'(s) \\ a(\Gamma(s)) & b(\Gamma(s)) & c(\Gamma(s)) \end{pmatrix} \leqslant 1, \qquad (7.2.14)$$

or, in other words,

$$\begin{vmatrix} f'(s) & g'(s) \\ a(f(s), g(s), h(s)) & b(f(s), g(s), h(s)) \end{vmatrix} =$$
$$\begin{vmatrix} f'(s) & h'(s) \\ a(f(s), g(s), h(s)) & c(f(s), g(s), h(s)) \end{vmatrix} =$$
$$\begin{vmatrix} g'(s) & h'(s) \\ b(f(s), g(s), h(s)) & c(f(s), g(s), h(s)) \end{vmatrix} = 0$$

for all $s \in I$.

A natural example of a smooth curve $\Gamma(s)$ with the property (7.2.14) is constructed as follows: take a characteristic curve $y(t)$ whose domain is an open interval J, and consider a C^1-invertible change of variables $t = \tau(s)$ in an open interval I. Then the curve $\Gamma(s) = y(\tau(s))$ on I satisfies the condition (7.2.14) everywhere on I:

$$\Gamma'(s) = (y(\tau(s)))' = \tau'(s)y'(\tau(s))$$
$$= \tau'(s)\mathcal{F}_{a,b,c}(y(\tau(s)))$$
$$= \tau'(s)\mathcal{F}_{a,b,c}(\Gamma(s))$$

for all $s \in I$.

Finally, we have to mention that whenever the determinant

$$\begin{vmatrix} f'(s) & g'(s) \\ a(f(s), g(s), h(s)) & b(f(s), g(s), h(s)) \end{vmatrix}$$

is equal to zero everywhere on the domain of a smooth curve Γ, then the curve need not satisfy (7.2.14).

Claim 7.2.6. *Following the notation of Theorem 7.2.2, if a smooth curve $\Gamma(s)$ is such that*

$$\begin{vmatrix} f'(s) & g'(s) \\ a(f(s), g(s), h(s)) & b(f(s), g(s), h(s)) \end{vmatrix} = 0$$

for all $s \in I$, then the corresponding Cauchy problem may have no solution, or infinitely many (pairwise nonequivalent) solutions.

Proof. Let us consider the (legitimate) first-order quasilinear PDE,

$$u_x = 0,$$

and a pair Cauchy problems involving it.

Writing down and solving the corresponding characteristic system, we obtain

$$\begin{cases} x'(t) = 1, \\ y'(t) = 0, \\ z'(t) = 0 \end{cases} \iff \begin{cases} x(t) = t + C_1, \\ y(t) = C_2, \\ z(t) = C_3, \end{cases}$$

where C_1, C_2, C_3 are arbitrary constants. In effect, the ranges of the characteristic curves are open intervals of the straight lines in $\mathbf{R}^3$ that are parallel to the x-axis.

(i) First, we consider a Cauchy problem *involving a characteristic curve*:

$$\begin{cases} u_x = 0, \\ u(s, 1) = 1, \quad (-\infty < s < +\infty); \end{cases}$$

the initial curve is then given by

$$\Gamma(s) = \begin{pmatrix} s \\ 1 \\ 1 \end{pmatrix}, \quad (s \in \mathbf{R})$$

and it is obviously characteristic. Now a function

$$u(x, y) = k(y),$$

where $k \in C^1(\mathbf{R})$, is a solution of the Cauchy problem on $\Omega = \mathbf{R}^2$ if and only $k(1) = 1$ (say, $u(x, y) = y^n$, where $n \in \mathbf{N}$). In effect, *there are infinitely many solutions of the corresponding Cauchy problem, which are pairwise nonequivalent.*

(ii) Next, consider the Cauchy problem

$$\begin{cases} u_x = 0, \\ u(s, 0) = s, \quad (0 < s < 1) \end{cases}$$

involving the curve

$$\Gamma(s) = \begin{pmatrix} f(s) \\ g(s) \\ h(s) \end{pmatrix} = \begin{pmatrix} s \\ 0 \\ s \end{pmatrix}, \quad (0 < s < 1)$$

whose tangent $\begin{pmatrix} 1 \\ 0 \\ 1 \end{pmatrix}$ *is never parallel to the field vector* $\begin{pmatrix} a \\ b \\ c \end{pmatrix} = \begin{pmatrix} 1 \\ 0 \\ 0 \end{pmatrix}$, *that is, never parallel the characteristic direction and for which*

$$\begin{vmatrix} f'(s) & g'(s) \\ a(f(s), g(s), h(s)) & b(f(s), g(s), h(s)) \end{vmatrix} = \begin{vmatrix} 1 & 0 \\ 1 & 0 \end{vmatrix} = 0$$

everywhere on its domain $(0, 1)$.

Then, given a domain Ω which contains the open interval $(0, 1) \times \{0\}$, *the corresponding Cauchy problem has no solution on Ω.* Indeed, suppose otherwise: let $u \in C^1(\Omega)$ be a solution of the Cauchy problem at hand on Ω, and let $s_0 \in (0, 1)$. Write P for the point $\begin{pmatrix} s_0 \\ 0 \end{pmatrix}$.

As $u_x = 0$ everywhere on Ω, we get, by Corollary 1.4.2, that there is an open neighborhood $O_P \subseteq \Omega$ of P such that

$$s = u(s, 0) = \text{const}$$

for infinitely many points $\begin{pmatrix} s \\ 0 \end{pmatrix} \in O_P$ satisfying $0 < s < 1$, which is absurd. $\qquad \square$

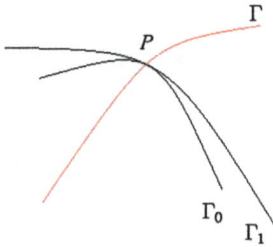

Figure 7.2.1: Creating infinitely many solutions of a Cauchy problem involving a characteristic curve (Γ), shown in red.

As we have seen above, there may be infinitely many pairwise nonequivalent solutions of a Cauchy problem involving a characteristic curve. It is, in fact, a quite "normal" situation, since the graphs of solutions of first-order quasilinear PDEs are made up of *characteristic* curves.

Indeed, consider a first-order quasilinear PDE and imagine its characteristic curve Γ and a family $\{\Gamma_n : n \in \mathbf{N}\}$ of smooth anticharacteristic curves with pairwise different ranges about 0, all meeting at the same point P on Γ at $t = 0$:

$$P = \Gamma_n(0), \quad (n \in \mathbf{N})$$

(see Figure 7.2.1). "Normally" then, one could expect that for every $n \in \mathbf{N}$, Γ (or, at least, its suitable part) is one of the characteristic curves out of which the graph of a solution u_n of the corresponding Cauchy problem involving the curve Γ_n is made. This will us give us infinitely many pairwise nonequivalent solutions u_n of the Cauchy problem involving the curve Γ.

Example 7.2.1. Use the method of characteristics to solve the Cauchy problem

$$\begin{cases} u_x + uu_y = -4, \\ u(-s, s) = 2s, \quad (s \in I), \end{cases}$$

where I is a maximal interval in $\mathbf{R}$ such that the transversality condition (7.2.10),

$$\begin{vmatrix} f'(s) & g'(s) \\ a(f(s), g(s), h(s)) & b(f(s), g(s), h(s)) \end{vmatrix} \neq 0,$$

associated with the problem, holds for all points $s \in I$.

Solution. Following the standard notation, we have that the coefficient functions of the PDE in question are the functions

$$a(x, y, z) = 1, \quad b(x, y, z) = z, \quad c(x, y, z) = -4$$

and the component functions defining the initial curve are the functions

$$f(s) = -s, \quad g(s) = s, \quad h(s) = 2s.$$

Then the transversality condition

$$\begin{vmatrix} f'(s) & g'(s) \\ a(f(s),g(s),h(s)) & b(f(s),g(s),h(s)) \end{vmatrix} \neq 0, \tag{7.2.15}$$

associated with our problem becomes

$$\begin{vmatrix} -1 & 1 \\ 1 & 2s \end{vmatrix} = -2s - 1 \neq 0,$$

or

$$s \neq -\frac{1}{2}.$$

Thus there are exactly two maximal intervals in **R** on which the condition (7.2.15) is true for all their points, namely, the open intervals

$$I_1 = (-\infty, -1/2) \quad \text{and} \quad I_2 = (-1/2, +\infty).$$

So we may work either with I_1, or with I_2 as I. At the moment, there is no need to make any choice, it is just enough to know what I can be.

We follow the main steps of the methods of characteristics. At the first step, given any $s \in I$, we have to solve the characteristic system

$$\begin{cases} x'(t) = a(x(t),y(t),z(t)), \\ y'(t) = b(x(t),y(t),z(t)), \\ z'(t) = c(x(t),y(t),z(t)), \end{cases}$$

subject to the initial conditions

$$x(0) = f(s), \quad y(0) = g(s), \quad z(0) = h(s).$$

In other words, we have to solve the system

$$\begin{cases} x'(t) = 1, \\ y'(t) = z(t), \\ z'(t) = -4, \end{cases}$$

subject to the initial conditions

$$x(0) = -s, \quad y(0) = s, \quad z(0) = 2s. \tag{7.2.16}$$

Let us obtain the general solution of the system. We derive from the third equation that

$$z(t) = -4t + C_3,$$

and, following this, derive from the second equation that

$$y'(t) = -4t + C_3 \Rightarrow y(t) = -2t^2 + C_3 t + C_2.$$

Further, the first equation implies that

$$x'(t) = 1 \Rightarrow x(t) = t + C_1.$$

Therefore,

$$\begin{cases} x(t) = t + C_1, \\ y(t) = -2t^2 + C_3 t + C_2, \\ z(t) = -4t + C_3, \end{cases}$$

where C_1, C_2, C_3 are arbitrary constants, and, as it is easy to see, the converse is also true. In particular,

$$\begin{cases} x(0) = C_1, \\ y(0) = C_2, \\ z(0) = C_3 \end{cases}$$

for any solution of the system. We therefore see that given any open interval J about 0, the solution which satisfies the initial conditions (7.2.16) is given by the functions

$$\begin{cases} x_s(t) = t - s, \\ y_s(t) = -2t^2 + 2st + s, \\ z_s(t) = -4t + 2s \end{cases}$$

on J.

Recall that the method requires a choice of a suitable open interval J_s about 0 for each $s \in I$, so that the set

$$\Theta = \bigcup_{s \in I} \{s\} \times J_s$$

is a domain in the st-plane and the function

$$\Psi \begin{pmatrix} s \\ t \end{pmatrix} = \begin{pmatrix} x_s(t) \\ y_s(t) \end{pmatrix}, \quad \begin{pmatrix} s \\ t \end{pmatrix} \in \Theta$$

is C^1-invertible.

Well, let us again leave open intervals J_s *unspecified* for a while, and try to figure out how to choose them so that the corresponding function Ψ on the corresponding set Θ is C^1-invertible. Define, following the notation of Theorem 7.2.2, the functions

$$X(s,t) = x_s(t), \quad Y(s,t) = y_s(t),$$
$$Z(s,t) = z_s(t)$$

on Θ, thereby getting that

$$X(s,t) = t - s, \quad Y(s,t) = -2t^2 + 2st + s,$$
$$Z(s,t) = -4t + 2s.$$

Further, the function

$$\Psi\begin{pmatrix} s \\ t \end{pmatrix} = \begin{pmatrix} X(s,t) \\ Y(s,t) \end{pmatrix} = \begin{pmatrix} t - s \\ -2t^2 + 2st + s \end{pmatrix}$$

must be C^1-invertible on our yet-to-be-constructed domain Θ, and so we must have that

$$\det \frac{\partial(X,Y)}{\partial(s,t)} = \begin{vmatrix} X_s(s,t) & X_t(s,t) \\ Y_s(s,t) & Y_t(s,t) \end{vmatrix} \neq 0$$

everywhere on Θ. It easily seen that

$$\det \frac{\partial(X,Y)}{\partial(s,t)} = \begin{vmatrix} -1 & 1 \\ 1 + 2t & 2s - 4t \end{vmatrix} = -2s + 2t - 1.$$

In effect, Θ must not contain the line

$$-2s + 2t - 1 = 0 \iff t - s = 1/2 \iff t = s + 1/2$$

To enhance the argument, the reader can verify that if Θ had met the line $t - s = 1/2$, we would have that

$$\Psi\begin{pmatrix} s \\ s + 1/2 \end{pmatrix} = \begin{pmatrix} 1/2 \\ -1/2 \end{pmatrix}$$

for infinitely many points of the form $\begin{pmatrix} s \\ s + 1/2 \end{pmatrix}$ in Θ, thereby demonstrating that Ψ would not have been 1-1, and hence invertible, on Θ.

This implies, keeping in mind that I is one of the intervals I_1, I_2 above, that we can run the method of characteristics either on the domain

$$\Theta = \Theta_1 = \bigcup_{s < -1/2} \{s\} \times (s + 1/2, +\infty)$$

when working with $I = I_1 = (-\infty, -1/2)$, or on the domain

$$\Theta = \Theta_2 = \bigcup_{s > -1/2} \{s\} \times (-\infty, s + 1/2)$$

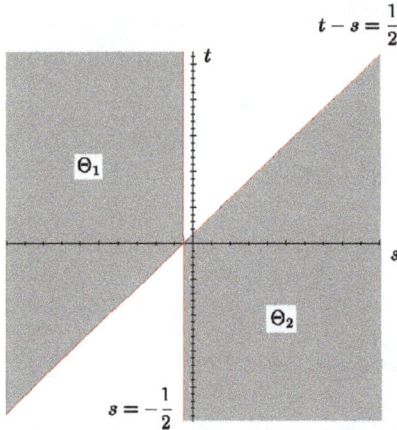

Figure 7.2.2: Domains Θ_1 and Θ_2, Example 7.2.1.

when working with $I = I_2 = (-1/2, +\infty)$ (see Figure 7.2.2). This settles the problem of choice of the intervals J_s. For instance, in the second case

$$J_s = (-\infty, s + 1/2)$$

for all $s \in I_2 = (-1/2, +\infty)$ (to remind the reader for the last time: we must make sure that $0 \in J_s$, right?).

Now having ensured that the corresponding function Ψ is invertible on Θ (equal either to Θ_1, or to Θ_2), let us find the inverse Ψ^{-1} of Ψ. Recall that this means, speaking in "plain" terms, obtaining expressions for s, t via x, y from the system

$$\begin{cases} x = X(s,t), \\ y = Y(s,t), \end{cases} \quad \text{or} \quad \begin{cases} x = t - s, \\ y = -2t^2 + 2st + s. \end{cases}$$

Let us move on, then, keeping in mind that, due to the above discussion,

$$t - s \neq 1/2 \iff x \neq 1/2$$

everywhere on Θ:

$$\begin{cases} x = t - s, \\ y = -2t^2 + 2st + s, \end{cases} \Rightarrow s = t - x$$

$$\Rightarrow y = -2t^2 + 2(t - x)t + (t - x)$$

$$\Rightarrow y = (1 - 2x)t - x$$

$$\Rightarrow x + y = (1 - 2x)t$$

whence

$$t = \frac{x + y}{1 - 2x}$$

and

$$s = t - x = \frac{2x^2 + y}{1 - 2x}.$$

This implies that

$$\Psi^{-1}\begin{pmatrix} x \\ y \end{pmatrix} = \begin{pmatrix} S(x,y) \\ T(x,y) \end{pmatrix}$$

on $\Omega = \Psi(\Theta)$, where

$$S(x,y) = \frac{2x^2 + y}{1 - 2x}$$

and

$$T(x,y) = \frac{x + y}{1 - 2x}$$

for all $\begin{pmatrix} x \\ y \end{pmatrix} \in \Omega$. Now since $x \neq 1/2$ for all points $\begin{pmatrix} x \\ y \end{pmatrix} \in \Omega$, the functions $S(x,y)$ and $T(x,y)$ are evidently continuously differentiable on Ω.

Finally, in "plain" terms the defining formula of the solution of the corresponding Cauchy problem on Ω is given by

$$z = -4t + 2s$$
$$= -4 \cdot \frac{x+y}{1-2x} + 2 \cdot \frac{2x^2 + y}{1-2x},$$
$$= \frac{4x^2 - 4x - 2y}{1 - 2x}.$$

Alternatively, to apply Theorem 7.2.2 in a more formal fashion, the solution we are looking for is the function

$$v(x,y) = Z(\, S(x,y), T(x,y)\,)$$
$$= -4T(x,y) + 2S(x,y)$$
$$= \frac{4x^2 - 4x - 2y}{1 - 2x}$$

on Ω. It is easy to verify that the function v is a solution of the PDE and satisfies the initial condition in question (please do so either by hand, or with the use of math software).

It may have been noticed that we have made no attempt to determine the domain $\Omega = \Psi(\Theta)$ explicitly (the reader will find it a pleasant exercise). But it is clearly unnecessary in view of the answer we have obtained. Indeed, what we need to know is that

Ω contains, by the construction, the range of the orthogonal projection of the initial curve Γ, and that it is a domain.

To complete consideration of the example: in light of our considerations above, if I is any interval in **R** such that $-1/2 \notin I$, a solution of the Cauchy problem

$$\begin{cases} u_x + uu_y = -4, \\ u(-s, s) = 2s, \quad (s \in I), \end{cases}$$

on any domain Ω in $\mathbf{R}^2$, which does not meet the line $x = 1/2$ and which contains the range of the curve $s \mapsto \begin{pmatrix} -s \\ s \end{pmatrix}$ on I, is given by the function

$$v(x, y) = \frac{4x^2 - 4x - 2y}{1 - 2x}$$

on Ω. □

Working on the last example, we took pains to construct explicitly the domain Θ in the st-plane to run the method of characteristics on. However, in practice, more often than not, such explicit constructions, as the reader may guess, are either very difficult, or plainly impossible, and/or plainly impracticable. A possible answer here may be to use of the *local* version of the method which allows us, keeping the corresponding notation, to work on some *unspecified* open rectangular neighborhood Θ of a point $\begin{pmatrix} s_0 \\ 0 \end{pmatrix}$, making along the way all reasonable assumptions on the functions $X(s, t)$ and $Y(s, t)$ to ensure invertibility of the map Ψ, knowing that they *will* hold if Θ is small enough, and so on. What is also important to note is that, quite often, a local solution associated with some $s_0 \in I$ the local version of the method produces is either perfectly fit to be a solution, or can be easily extended to a solution of the corresponding Cauchy problem (for the whole of I, or even beyond). In a word, the local version of the method of characteristics often delivers more than promises.

Example 7.2.2. Consider the Cauchy problem

$$\begin{cases} uu_x + yu_y = x, \\ u(0, s) = s, \quad (s \in I), \end{cases}$$

where I is a maximal interval all whose points satisfy the transversality condition (7.2.10). Find local solutions of the Cauchy problem for all points s_0 of the interval I, and consider the problem of extending them to appropriate domains in $\mathbf{R}^2$.

Solution. We shall work with Maple, discussing, along the way, possible problems in need of handling on similar occasions.

The following is a template for the (local version of the) method of characteristics. The template is constructed along the lines of Theorem 7.2.2. We suggest the reader to study the template carefully.

```
restart;

with(linalg):

### Coefficients

a := (x,y,z) -> ... :
b := (x,y,z) -> ... :
c := (x,y,z) -> ... :

###  Initial Curve

f := s -> ... :
g := s -> ... :
h := s -> ... :

Gamma := s -> (f(s), g(s), h(s)) :

### Transversality condition

DET := det( matrix(2,2,
  [ diff( f(s),s ), diff( g(s),s ),
  a(Gamma(s)), b(Gamma(s)) ])
);

### Characteristic System

sys:=diff( x(t),t ) = a( x(t),y(t),z(t) ),
  diff( y(t),t ) = b( x(t),y(t),z(t) ),
  diff( z(t),t ) = c( x(t),y(t),z(t) ):

### Characteristic curves gamma_s(t)

dsolve(
  { sys, x(0)=f(s), y(0)=g(s), z(0)=h(s) },
  { x(t),y(t),z(t) } );

### Functions X(s,t), Y(s,t), Z(s,t)

#2#    X := (s,t) -> ... :
#2#    Y := (s,t) -> ... :
#2#    Z := (s,t) -> ... :
```

```
          ###

#2#       solve( {x=X(s,t), y=Y(s,t)},{s,t});

          ### Components of the Inverse of Psi:
          ###     functions S(x,y), T(x,y)

#3#       S := (x,y) -> ... :
#3#       T := (x,y) -> ... :


          ### Solution

#3#       u := (x,y) -> Z( S(x,y), T(x,y) ):

#3#       u(x,y);

#3#       print(`Solution `,simplify( u(x,y) ) );

          ### Solution of the PDE ?

#3#       simplify(
#3#           a(x,y,u(x,y))*diff( u(x,y),x ) +
#3#           b( x,y,u(x,y) )*diff( u(x,y),y )
#3#           -c( x,y,u(x,y) )
#3#       );

          ### Initial conditions ?

#3#       simplify( u( f(s),g(s) ) -h(s) );
```

One can use the above template in much the same way as we used the previous template (when working on Example 6.4.2). At the first step, enter the initial data, and execute the program. Once new data arrives, remove the characters

> #2#

at the beginnings of the lines that follow, replace dots with relevant input, and run the program for the second time. Following this, remove the symbols

> #3#

enter relevant data again, and run the program for the third and the final time.

So we start with entering the initial data:

Coefficients

```
a := (x,y,z) -> z :
b := (x,y,z) -> y :
c := (x,y,z) -> x :
```

Initial Curve

```
f := s -> 0 :
g := s -> s :
h := s -> s :
```

Maple responds with

$$DET := -s$$

$$\left\{ y(t) = e^t s, \right.$$

$$x(t) = \left(\frac{1}{2}e^t - \frac{1}{2}e^{(-t)} \right)s,$$

$$\left. z(t) = \left(\frac{1}{2}e^t + \frac{1}{2}e^{(-t)} \right)s \right\}$$

This means that the transversality condition (7.2.10) becomes the condition

$$-s \neq 0.$$

Accordingly, take a real number $s_0 \neq 0$. The second line of the output implies, keeping notation of Theorem 7.2.2, that the defining functions X, Y, and Z of the Monge surface on an open neighborhood Θ of the point $\begin{pmatrix} s_0 \\ 0 \end{pmatrix}$ are

$$X(s,t) = s\frac{e^t - e^{-t}}{2} = s\sinh(t),$$

$$Y(s,t) = se^t,$$

$$Z(s,t) = s\frac{e^t + e^{-t}}{2} = s\cosh(t)$$

for all $\begin{pmatrix} s \\ t \end{pmatrix} \in \Theta.$

Working with a math software, one must always try to interpret the results the software provides as, well, a professional mathematician would. For, in fact, we have a local solution (almost) ready. Indeed, having the Monge surface over Θ presented as

$$\begin{cases} x = s\dfrac{e^t - e^{-t}}{2}, \\ y = se^t, \\ z = s\dfrac{e^t + e^{-t}}{2}, \end{cases} \tag{7.2.17}$$

we see that

$$z = y - x,$$

which means the Monge surface is (locally) on the plane

$$z = y - x,$$

and (7.2.17) is a bit unusual (local) parametrization of this plane.

Recall that, in "plain" terms, our goal when applying the method of characteristics is to derive from the equations of the Monge surface a formula

$$z = u(x, y),$$

expressing z in terms of x and y. We have accomplished just that, by observing that

$$\begin{cases} x = s\dfrac{e^t - e^{-t}}{2}, \\ y = se^t, \\ z = s\dfrac{e^t - e^{-t}}{2}, \end{cases} \Rightarrow z = y - x,$$

and so for every $s_0 \neq 0$, a local solution of the Cauchy problem is given by the function

$$u(x, y) = y - x.$$

Understandably, opportunities like that are quite rare in practice, and so should not be missed.

We postpone the formal conclusion of the example. Suppose, we did miss a good chance to complete the solution, and let us continue the solution with Maple, as if nothing happened.

We enter the second portion of data, thereby getting

```
### Functions X(s,t), Y(s,t), Z(s,t)

# x(t) = (1/2*exp(t)-1/2*exp(-t))*s,
# y(t) = exp(t)*s,
# z(t) = (1/2*exp(-t)+1/2*exp(t))*s

X := (s,t) -> (1/2*exp(t)-1/2*exp(-t))*s :
Y := (s,t) -> exp(t)*s :
Z := (s,t) -> (1/2*exp(-t)+1/2*exp(t))*s :

###

solve( {x=X(s,t), y=Y(s,t)},{s,t} );
```

the reader may have observed that we copied the solution of the characteristic system from the last output, and defined the function X, Y, and Z accordingly, by doing simple copy/pastes.

Let us run the program for the second time. Maple will add the line

$$\left\{ s = \frac{y}{\text{RootOf}((2x - y)\_Z^2 + y)}, \right.$$

$$\left. t = \ln(\text{RootOf}((2x - y)\_Z^2 + y)) \right\}$$

to the above output. We have already seen that Maple uses symbol(s) like $\_Z$ to refer to solutions of functional equations. In our case, $\_Z$ refers to a solution of the functional equation

$$(2x - y)\_Z^2 + y = 0.$$

As it is, there is more than one variant for

$$\text{RootOf}((2x - y)\_Z^2 + y),$$

and so we have to make an additional assumption on s_0: we have assumed earlier that $s_0 \neq 0$, and so let us make a stronger assumption that

$$s_0 > 0.$$

Under this assumption, we may further assume that Θ, an open disk/rectangle containing the point $\begin{pmatrix} s_0 \\ 0 \end{pmatrix}$ on which it is possible to run the method of characteristics, is such that

$$\begin{pmatrix} s \\ t \end{pmatrix} \in \Theta \Rightarrow s > 0.$$

Now the map

$$\Phi\begin{pmatrix} s \\ t \end{pmatrix} = \begin{pmatrix} X(s,t) \\ Y(s,t) \end{pmatrix} = \begin{pmatrix} s\sinh(t) \\ se^t \end{pmatrix} = \begin{pmatrix} x \\ y \end{pmatrix}$$

on Θ takes a point $\begin{pmatrix} s \\ t \end{pmatrix} \in \Theta$ to a point $\begin{pmatrix} x \\ y \end{pmatrix}$ in the xy-plane such that

$$y = se^t > 0.$$

Next, we have that

$$y - 2x = se^t - 2 \cdot s\frac{e^t - e^{-t}}{2} = se^{-t} > 0.$$

So to resolve uncertainty in

$$s = \frac{y}{\text{RootOf}((2x - y)\_Z^2 + y)},$$

we have to take as a function

$$\text{RootOf}((2x - y)\_Z^2 + y)$$

the function

$$\sqrt{\frac{y}{y - 2x}},$$

whence

$$s = \sqrt{y(y - 2x)}.$$

Accordingly,

$$t = \ln(\text{RootOf}((2x - y)\_Z^2 + y))$$

must be

$$t = \ln\sqrt{\frac{y}{y - 2x}}.$$

We are in a position to enter the last bunch of data:

```
assume( s > 0, y > 0 , y-2*x > 0 ) :

S := (x,y) -> sqrt( y*(y-2*x) ) :
T := (x,y) -> 1/2*ln( y/(y-2*x) ) :
```

Note the natural-looking command

```
assume( s > 0, y > 0 , y-2*x > 0 ) :
```

which will help Maple to cope with the task. We uncomment all other lines which begin with #3#, and run the program for the third (and the final) time.

Maple will add to the previous output a new line describing the (local) solution of the Cauchy problem and two more lines

$$Solution, -x^{\sim} + y^{\sim}$$
$$0$$
$$0$$

which apparently means that the function

$$u(x, y) = -x + y$$

is a required solution, and that this function satisfies the PDE in question, and meets the initial conditions. The symbols like

$$x^{\sim}, y^{\sim}, \ldots$$

imply that some restrictions are imposed on the corresponding variables (and truly, due to the command `assume(...)` we have added to the program).

We suggest the reader to figure out what must be changed in the program to investigate the case when $s_0 < 0$. Having done this, the reader will see that the method produces a function with the same defining formula. Alternatively, though ...?

Now we can summarize the above considerations at last: *formally*, by Proposition 7.2.5, if I is any interval in **R** satisfying $0 \notin I$, then a *local* solution of the Cauchy problem

$$\begin{cases} uu_x + yu_y = x, \\ u(0, s) = s, \quad (s \in I) \end{cases} \tag{7.2.18}$$

at any $s_0 \in I$ is given by the function

$$v(x, y) = y - x.$$

However, recalling our comment above on "delivering more than promised," as the function $v(x, y) = y - x$ is defined everywhere on $\mathbf{R}^2$, it is obviously a solution of the Cauchy problem in question for *any* interval I in **R** and *any* domain of Ω in $\mathbf{R}^2$ which contains the range $\{0\} \times I$ of the base curve. $\qquad\square$

Exercises

1. Use the method of characteristics to solve the Cauchy problem

$$\begin{cases} u_x - uu_y = 1, \\ u(-6s, 8s) = -3s, \quad (s \in I), \end{cases}$$

where I is a maximal interval in $\mathbf{R}$ such that the transversality condition (7.2.10), associated with the problem, holds for all points $s \in I$.

2. Consider the Cauchy problem

$$\begin{cases} (y + u)u_x - 4yu_y = x - y, \\ u(s, 7) = s, \quad (s \in I), \end{cases}$$

where I is a maximal interval all whose points satisfy the transversality condition (7.2.10). Find local solutions of the Cauchy problem for all points s_0 of the interval I, and consider the problem of extending them to appropriate domains in $\mathbf{R}^2$. (The reader must have the general solution of the characteristic system handy, provided that Exercise 5.2 from Chapter 5 has been solved.)

3. Use Maple and the program template given in Example 7.2.2 to obtain local solutions of Cauchy problems at suitable points of the domains of the initial curves:

(i) $\quad \begin{cases} uu_x - uu_y = 1, \\ u(s, s) = -s, \quad (s \in I); \end{cases}$

(ii) $\quad \begin{cases} (x + u)u_x + (y + u)u_y = u, \\ u(s, 0) = -s, \quad (s \in J), \end{cases}$

where I, J are open intervals in $\mathbf{R}$.

8 Second-order semilinear equations

8.1 Three main types

Definition (Second-order semilinear PDEs). A second-order PDE is called *semilinear* if it can be written in the form

$$\underbrace{a(x,y)u_{xx} + 2b(x,y)u_{xy} + c(x,y)u_{yy}}_{\text{The principal part}} = d(x,y,u,u_x,u_y). \tag{8.1.1}$$

The function

$$\Delta = \Delta(x,y) = b(x,y)^2 - a(x,y)c(x,y)$$
$$= b^2 - ac$$

is called the *discriminant* of the PDE (8.1.1). △

As above, when defining the discriminant, we will usually omit reference to the variables in the coefficient functions $a(x,y)$, $b(x,y)$, $c(x,y)$ in the principal part:

$$\underbrace{au_{xx} + 2bu_{xy} + cu_{yy}}_{\text{The principal part}} = d(x,y,u,u_x,u_y).$$

A matter of importance: notice that the coefficient of the partial derivative u_{xy} of a second-order semilinear PDE (8.1.1) is written as a multiple of 2. This makes these PDEs easier to deal with.

Recall that, quite similarly, the general equation

$$a_{11}x^2 + 2a_{12}xy + a_{22}y^2 + 2a_{13}x + 2a_{23}y + a_{33} = 0$$

of a second-order plane curve (a conic section, or simply a conic, an object of study in analytic geometry), the coefficient of xy is also written as a multiple of two.

As we have seen in Chapter 4, if the coefficient functions of a first-order semilinear PDE

$$g(x,y)u_x + h(x,y)u_y = k(x,y,u),$$

the functions $g(x,y)$ and $h(x,y)$, satisfy the conditions of the method of characteristics for first-order semilinear equations (Proposition 4.2.3), then there is a change of variables

$$\begin{cases} s = x, \\ t = T(x,y) \end{cases} \iff \begin{cases} x = s, \\ y = Y(s,t), \end{cases}$$

https://doi.org/10.1515/9783110677256-008

which transforms the original equation to the equation

$$g(s, Y(s,t))w_s = k(s, Y(s,t), w),$$

where $w(s,t) = u(x,y)$.

Knowing that, one may wonder is there a suitable change of variables, which can transform/reduce a *second*-order semilinear PDE to a simpler one. As the reader will see in due course, the answer to this question is affirmative (under "normal" circumstances). First, we will study the effect of a change of variables on the principal part of a second-order semilinear PDE.

Proposition 8.1.1. *Consider a second-order semilinear PDE*

$$au_{xx} + 2bu_{xy} + c(x,y)u_{yy} = d(x,y,u,u_x,u_y)$$

on a domain Ω in $\mathbf{R}^2$, where $a, b, c \in C(\Omega)$ and $d \in C(\Omega \times \mathbf{R}^3)$, and a change of variables

$$\begin{cases} s = S(x,y), \\ t = T(x,y) \end{cases}$$

which is C^1-invertible on Ω. Then if

$$Aw_{ss} + 2Bw_{st} + Cw_{tt} = D(s, t, w, w_s, w_t),$$

where $w(s,t) = u(x,y)$, is the corresponding transformed equation, the coefficients of its principal part are as follows:

$$A = as_x^2 + 2bs_x s_y + cs_y^2,$$
$$B = as_x t_x + b(s_x t_y + s_y t_x) + cs_y t_y,$$
$$C = at_x^2 + 2bt_x t_y + ct_y^2.$$

Furthermore,

$$B^2 - AC = (b^2 - ac)\left(\det \frac{\partial(s,t)}{\partial(x,y)} \right)^2, \tag{8.1.2}$$

where

$$\frac{\partial(s,t)}{\partial(x,y)} = \begin{pmatrix} s_x & s_y \\ t_x & t_y \end{pmatrix}$$

is the Jacobian matrix of the change of variables $x, y \to s, t$.

In effect, if the sign of the "old" discriminant $b^2 - ac$ is constant on Ω, then the sign of the "new" discriminant $B^2 - AC$ is also constant on the domain $\Phi(\Omega)$, where

$$\Phi\begin{pmatrix} x \\ y \end{pmatrix} = \begin{pmatrix} S(x,y) \\ T(x,y) \end{pmatrix}, \quad \begin{pmatrix} x \\ y \end{pmatrix} \in \Omega,$$

and is equal to the sign of the "old" discriminant.

Proof. Let $u \in C^1(\Omega)$ be a continuously differentiable function.

As we have to know, we start with the identity

$$u(x,y) = w(s,t) \quad // = w(S(x,y), T(x,y)) \quad //$$

on Ω, where w is a continuously differentiable function, whose existence due to the fact that our change of variables is C^1-invertible, and then differentiate this identity by x and by y with the use of the chain rule:

$$u_x = w_s s_x + w_t t_x \quad \text{and} \quad u_y = w_s s_y + w_t t_y.$$

Then, for instance, to get the formula/identity for u_{xy}, we further differentiate, say, the first identity

$$u_x = w_s s_x + w_t t_x$$

by y:

$$
\begin{aligned}
u_{xy} &= [w_s s_x + w_t t_x]_y \\
&= [w_s s_x]_y + [w_t t_x]_y \\
&= [w_s]_y s_x + w_s [s_x]_y + [w_t]_y t_x + w_t [t_x]_y \\
&= (w_{ss} s_y + w_{st} t_y) s_x + w_s s_{xy} + (w_{ts} s_y + w_{tt} t_y) t_x + w_t t_{xy} \\
&= s_x s_y w_{ss} + (s_x t_y + s_y t_x) w_{st} + t_x t_y w_{tt} + s_{xy} w_s + t_{xy} w_t.
\end{aligned}
$$

Working in a similar fashion, we obtain that

$$u_{xx} = s_x^2 w_{ss} + 2 s_x t_x w_{st} + t_{xx}^2 w_{tt} + s_{xx} w_s + t_{xx} w_t,$$

and, *by symmetry* (replacing x with y in the last identity),

$$u_{yy} = s_y^2 w_{ss} + 2 s_y t_y w_{st} + t_{yy}^2 w_{tt} + s_{yy} w_s + t_{yy} w_t.$$

Now, putting everything together, we obtain

$$
\begin{aligned}
a u_{xx} &= a(s_x^2 w_{ss} + 2 s_x t_x w_{st} + t_{xx}^2 w_{tt} + s_{xx} w_s + t_{xx} w_t), \\
2 b u_{xy} &= 2b(s_x s_y w_{ss} + (s_x t_y + s_y t_x) w_{st} + t_x t_y w_{tt} + s_{xy} w_s + t_{xy} w_t), \\
c u_{yy} &= c(s_y^2 w_{ss} + 2 s_y t_y w_{st} + t_{yy}^2 w_{tt} s_{yy} w_s + t_{yy} w_t)
\end{aligned}
\tag{8.1.3}
$$

and the result quickly follows:

$$
\begin{aligned}
& au_{xx} + 2bu_{xy} + cu_{yy} \\
&= \underbrace{\left(as_x^2 + 2bs_xs_y + cs_y^2\right)}_{A} w_{ss} + \cdots \\
&= Aw_{ss} + 2\underbrace{\left(as_xt_x + b(s_xt_y + s_yt_x) + cs_yt_y\right)}_{B} w_{st} + \cdots \\
&= Aw_{ss} + 2Bw_{st} + \underbrace{\left(at_x^2 + 2bt_xt_y + ct_y^2\right)}_{C} w_{tt} + \cdots \\
&= Aw_{ss} + 2Bw_{st} + Cw_{tt} + \cdots
\end{aligned}
$$

It is easy to see that

$$
\begin{pmatrix} A & B \\ B & C \end{pmatrix} = \begin{pmatrix} s_x & s_y \\ t_x & t_y \end{pmatrix} \begin{pmatrix} a & b \\ b & c \end{pmatrix} \begin{pmatrix} s_x & s_y \\ t_x & t_y \end{pmatrix}^t ,
$$

where M^t denote the transpose of a matrix M (one could guess that the last formula is true, knowing that a similar result holds for the matrices of coefficients of a given conic section and an affine change of variables). The reader can use the following program in Maple to verify the above formula:

```
restart;
with(linalg):

A := a*s_x^2 + 2*b*s_x*s_y +c* s_y^2:
B := a*s_x*t_x + b*(s_x*t_y+ s_y*t_x) +c*s_y*t_y:
C := a*t_x^2 + 2*b*t_x*t_y +c*t_y^2:

M1:=matrix(2,2,[A,B,B,C]):
M:=matrix(2,2,[a,b,b,c]):
J:=matrix(2,2,[s_x,s_y,t_x,t_y]):

X:=evalm( M1 - J &* M &* transpose(J) ):
expand( X[1,1] );
expand( X[1,2] );
expand( X[2,1] );
expand( X[2,2] );
```

Consequently,

$$
\det \begin{pmatrix} A & B \\ B & C \end{pmatrix}
$$

$$= \det\left(\begin{pmatrix} s_x & s_y \\ t_x & t_y \end{pmatrix}\begin{pmatrix} a & b \\ b & c \end{pmatrix}\begin{pmatrix} s_x & s_y \\ t_x & t_y \end{pmatrix}^t\right)$$

$$= \det\begin{pmatrix} s_x & s_y \\ t_x & t_y \end{pmatrix}\det\begin{pmatrix} a & b \\ b & c \end{pmatrix}\det\begin{pmatrix} s_x & s_y \\ t_x & t_y \end{pmatrix}^t$$

$$= \det\begin{pmatrix} a & b \\ b & c \end{pmatrix}\det\begin{pmatrix} s_x & s_y \\ t_x & t_y \end{pmatrix}^2,$$

whence

$$B^2 - AC = (b^2 - ac)\left(\det\frac{\partial(s,t)}{\partial(x,y)}\right)^2,$$

as claimed. ☐

Agreement. Everywhere in this chapter, if a second-order semilinear PDE

$$au_{xx} + 2bu_{xy} + cu_{yy} = d(x,y,u,u_x,u_y)$$

is defined on an open set Ω in $\mathbf{R}^2$, we shall always assume, as it is done in Proposition 8.1.1, that the coefficient functions

$$a = a(x,y), \quad b = b(x,y), \quad c = c(x,y)$$

are continuous on Ω, and that the function $d = d(x,y,z,v,w)$ is continuous on $\Omega \times \mathbf{R}^3$. $\triangle$

Definition (Types of second-order semilinear PDEs). Let Ω be an open set in $\mathbf{R}^2$ and let

$$au_{xx} + 2bu_{xy} + cu_{yy} = d(x,y,u,u_x,u_y) \tag{8.1.4}$$

be a second-order semilinear PDE on Ω. Then
(i) the equation (8.1.4) is *hyperbolic* on Ω if

$$b^2 - ac > 0$$

everywhere on Ω, is *parabolic* on Ω if

$$b^2 - ac = 0$$

everywhere on Ω, and is *elliptic* on Ω if

$$b^2 - ac < 0$$

everywhere on Ω.
(ii) if P is a point of Ω and if the discriminant of the equation (8.1.4) is of constant sign on some open neighborhood O_P of P, then the equation (8.1.4) is said to be *hyperbolic* (resp., *parabolic*, resp., *elliptic*) about P if the sign of the discriminant is positive (resp., equal to zero, resp., negative) everywhere on O_P. $\triangle$

Summarizing Proposition 8.1.1, we can say that the type of a second-order semilinear PDE on a certain open set (if any) remains the same under a C^1-invertible change of variables.

A terminological remark: the terms "hyperbolic," "elliptic," and "parabolic" are chosen to reflect on the analogy with the affine classification of conics (plane curves of degree two). One of the key facts which are crucial to the said classification is that the discriminant

$$a_{12}^2 - a_{11}a_{22}$$

a conic section with the equation

$$a_{11}x^2 + 2a_{12}xy + a_{22}y^2 + 2a_{13}x + 2a_{23}y + a_{33} = 0$$

is of the same sign in any affine coordinate system (keeps the same sign under any affine change of variables/coordinates; we have already referred to the proof this result above).

Recall that given any hyperbola in the plane $\mathbf{R}^2$, there is an affine coordinate system (an affine change of variables) with respect to which the equation of the hyperbola becomes

$$x^2 - y^2 = 1, \quad \text{or} \quad xy = 1,$$

and it is called the *canonical* equation of a hyperbola; accordingly, the discriminant of any hyperbola is positive. Likewise, the canonical equation of an ellipse with respect to a suitable affine coordinate system of $\mathbf{R}^2$ is

$$x^2 + y^2 = 1,$$

and hence the discriminant of any ellipse is negative, and the canonical affine equation of a parabola in the plane is

$$y^2 = x,$$

and its discriminant is equal to zero.

Examples

(i) (a) Equations of the form

$$u_{xy} = d(x, y, u, u_x, u_y)$$

and

$$u_{xx} - u_{yy} = d(x, y, u, u_x, u_y)$$

are *hyperbolic*:

$$\Delta_1 = \frac{1}{4} - 0 \cdot 0 > 0 \quad \text{and}$$

$$\Delta_2 = 0^2 - 1 \cdot (-1) = 1 > 0$$

(cf. the canonical equations

$$xy = 1 \quad \text{and} \quad x^2 - y^2 = 1$$

of hyperbolas).

   (b) Equations of the form

$$u_{xx} = d(x, y, u, u_x, u_y)$$

and

$$u_{yy} = d(x, y, u, u_x, u_y)$$

are *parabolic*:

$$\Delta_1 = \Delta_2 = 0^2 - 1 \cdot 0 = 0^2 - 0 \cdot 1 = 0$$

(cf. the canonical equation $y^2 = x$ of parabolas).

   (c) Equations of the form

$$u_{xx} + u_{yy} = d(x, y, u, u_x, u_y)$$

are elliptic:

$$\Delta = 0^2 - 1 \cdot 1 = -1 < 0$$

(cf. . . . .).

   Now some classical examples from physics.

   (ii) The *wave equation*

$$u_{tt} - c_w u_{xx} = 0,$$

where $c_w > 0$ is a positive constant (the speed of the waves), is hyperbolic on $\mathbf{R}^2$:

$$a = 1, b = 0, c = -c_w < 0 \Rightarrow b^2 - ac = c_w > 0.$$

(iii) The *heat conduction equation*

$$u_t - a^2 u_{xx} = 0,$$

where $a > 0$, is parabolic on $\mathbf{R}^2$.

(iv) *Laplace's equation*

$$u_{xx} + u_{yy} = 0$$

is elliptic.

(v) The type of a PDE on an open set Ω is a local property, that is, understandably, it depends on the choice of Ω. So a given PDE can have different type in different regions in $\mathbf{R}^2$. Furthermore, a PDE can contain parameters, and so its type may change as parameters change.

For instance, the *Euler–Tricomi* equation

$$yu_{xx} + u_{yy} = 0, \quad (b^2 - ac = 0 - y = -y)$$

is elliptic in the upper open half-plane $\{y > 0\}$ and hyperbolic in the lower open half-plane $\{y < 0\}$.

An important equation

$$(1 - M^2)u_{xx} + u_{yy} = 0, \quad (b^2 - ac = M^2 - 1)$$

from continuum mechanics, where $M > 0$ is a positive constant, is elliptic if $M < 1$, hyperbolic if $M > 1$, and parabolic if $M = 1$.

Now suppose we have performed a C^1-invertible change of variables

$$s = s(x, y), \quad t = t(x, y)$$

in a second-order semilinear PDE on a certain domain Ω, thereby getting a transformed PDE

$$Aw_{ss} + 2Bw_{st} + Cw_{tt} = E(s, t, w, w_s, w_t).$$

Let us think under what conditions on the functions s, t, the coefficient

$$A = as_x^2 + 2bs_xs_y + cs_y^2$$

will disappear (or, in other words, the corresponding function $A = A(s, t)$ will become identically zero on the new domain).

How to achieve that? Well, suppose that $s_y \neq 0$ everywhere on Ω. Then

$$A = s_y^2 \left(a \left(\frac{s_x}{s_y} \right)^2 + 2b \left(\frac{s_x}{s_y} \right) + c \right).$$

We therefore conclude that if the function

$$\frac{s_x}{s_y}$$

is one of the continuous functions D satisfying the functional equation

$$aD^2 + 2bD + c = 0$$

identically on Ω, we shall achieve our goal.

Suppose further that $a \neq 0$ everywhere on Ω. Then for any real-valued function D on Ω satisfying

$$aD^2 + 2bD + c = 0$$

identically on Ω, we have that

$$D = \frac{-b \pm \sqrt{b^2 - ac}}{a},$$

where, necessarily, $b^2 - ac \geqslant 0$ everywhere on Ω.

However, if we require D be *continuous*, then either

$$D = D_1 = \frac{-b + \sqrt{b^2 - ac}}{a},$$

or

$$D = D_2 = \frac{-b - \sqrt{b^2 - ac}}{a}$$

everywhere on Ω.

Similarly, if we wish to "kill" the coefficient C, we, under the assumption that $t_y \neq 0$ everywhere on Ω, have to have, that

$$\frac{t_x}{t_y} = D_1, \quad \text{or} \quad \frac{t_x}{t_y} = D_2, \tag{8.1.5}$$

everywhere on Ω. Fortunately, as we shall below, situations like that are quite possible in practice (spoiler: if a function $t = t(x,y)$ satisfies (8.1.5), then it is a solution of one of the first-order linear PDEs $t_x - D_{1,2}t_y = 0$).

OK, given a second-order linear PDE on a domain Ω with the coefficients

$$a = a(x,y), \quad b = b(x,y), \quad c = c(x,y)$$

in the principal part, we define the functions

$$D_1 = \frac{-b + \sqrt{b^2 - ac}}{a}, \quad \text{and} \quad D_2 = \frac{-b - \sqrt{b^2 - ac}}{a}$$

on Ω provided that the discriminant is nonnegative everywhere on Ω and the function a is nowhere zero on Ω. Observe that we then have, by Vieta's theorem, that

$$D_1 + D_2 = -\frac{2b}{a} \quad \text{and} \quad D_1 D_2 = \frac{c}{a}. \tag{8.1.6}$$

8.2 Hyperbolic equations

Proposition 8.2.1 (Canonical form of the hyperbolic equations). *Let*

$$au_{xx} + 2bu_{xy} + cu_{yy} = d(x,y,u,u_x,u_y) \tag{8.2.1}$$

be a second-order semilinear equation on a domain Ω in $\mathbf{R}^2$, let the function a be nonzero everywhere on Ω, and let

$$b^2 - ac > 0$$

everywhere on Ω. Let further the functions

$$D_1 = \frac{-b + \sqrt{b^2 - ac}}{a}$$

and

$$D_2 = \frac{-b - \sqrt{b^2 - ac}}{a}$$

be continuously differentiable on Ω.

(i) Suppose that for a point P of Ω there exists a subdomain Ω_P of Ω, which contains P, such that
   (a) both the ODE

$$y' = -D_1$$

and the ODE

$$y' = -D_2$$

possess first integrals $s(x,y)$ and $t(x,y)$, respectively, on Ω_P;

(b) *both the partial derivative* s_y *and the partial derivative* t_y *are nonzero everywhere on* Ω_P.

Then there is a subdomain Π_P of Ω which contains the point P such that the change of variables

$$\begin{cases} s = s(x,y), \\ t = t(x,y), \end{cases} \tag{8.2.2}$$

reduces the PDE (8.2.1) *on* Π_P *to a PDE of the form*

$$w_{st} = d^*(s,t,w,w_s,w_t), \tag{8.2.3}$$

called the first canonical form *of the hyperbolic PDEs.*

(ii) *Accordingly, if both the ODEs*

$$y' = -D_1 \quad and \quad y' = -D_2$$

have first integrals $s(x,y)$ *and* $t(x,y)$, *respectively, on the whole of* Ω *such that the change of variables* (8.2.3) *is* C^1-*invertible on* Ω, *then under this change of variables the PDE* (8.2.1) *on* Ω *is reduced to a PDE in the canonical form* (8.2.3).

Proof. (i) First of all, due to being, by (a), a first integral of the ODE

$$y' = -D_1,$$

on Ω_P, the function s satisfies, by Proposition 4.1.1, the PDE

$$U_x - D_1 U_y = 0$$

on Ω_P, and hence

$$s_x - D_1 s_y = 0$$

everywhere on Ω_P. Consequently, by (b),

$$D_1 = \frac{s_x}{s_y}$$

at all points of Ω. Similarly,

$$D_2 = \frac{t_x}{t_y}$$

everywhere on Ω_P.

As we have remarked above, the coefficient A of the transformed equation under the change of variables $x, y \to s, t$ is then equal to zero (provided that this change of variables is C^1-invertible). Let us repeat the argument anyway:

$$A = s_y^2 \left(a \left(\frac{s_x}{s_y} \right)^2 + 2b \left(\frac{s_x}{s_y} \right) + c \right)$$
$$= s_y^2 (aD_1^2 + 2bD_1 + c)$$
$$= 0.$$

Similarly, $C = 0$.

But $B \neq 0$. Indeed, by (8.1.6),

$$B = as_x t_x + b(s_x t_y + s_y t_x) + cs_y t_y$$
$$= s_y t_y \left(a \cdot \frac{s_x}{s_y} \cdot \frac{t_x}{t_y} + b \left(\frac{s_x}{s_y} + \frac{t_x}{t_y} \right) + c \right)$$
$$= s_y t_y (aD_1 D_2 + b(D_1 + D_2) + c)$$
$$= s_y t_y \left(a \cdot \frac{c}{a} + b \cdot -\frac{2b}{a} + c \right)$$
$$= \frac{2}{a} \cdot s_y t_y \cdot (ac - b^2) \neq 0$$

everywhere on Ω_P.

Finally, by the Inverse Function Theorem, the change of variables

$$\begin{cases} s = s(x, y), \\ t = t(x, y) \end{cases}$$

is C^1-invertible on a suitable subdomain $\Pi_P \subseteq \Omega_P$, which contains the point P. Indeed,

$$\det \frac{\partial(s, t)}{\partial(x, y)} = \begin{vmatrix} s_x & s_y \\ t_x & t_y \end{vmatrix} = s_y t_y \begin{vmatrix} s_x/s_y & 1 \\ t_x/t_y & 1 \end{vmatrix} \tag{8.2.4}$$
$$= s_y t_y \begin{vmatrix} D_1 & 1 \\ D_2 & 1 \end{vmatrix} = s_y t_y (D_1 - D_2),$$

at every point of Ω_P, and so the Jacobian determinant is nonzero due to the fact that

$$s_y \neq 0, \quad t_y \neq 0, \quad \text{and} \quad D_1 \neq D_2$$

everywhere on Ω_P.

(ii) The result easily follows, by (i), provided that

$$s_y(P) \neq 0 \quad \text{and} \quad t_y(P) \neq 0,$$

at all points P of Ω. Suppose then, for instance, that $s_y(P) = 0$ at some point $P \in \Omega$. However, as the function s is, by Proposition 4.1.1, a solution of the PDE

$$U_x - D_1 U_y = 0 \quad (\Rightarrow s_x - D_1 s_y = 0)$$

on Ω, then $s_x(P) = 0$ as well, and the Jacobian determinant of the change of variables $x, y \to s, t$ is equal to zero at P,

$$\det \frac{\partial(s,t)}{\partial(x,y)}(P) = \begin{vmatrix} s_x(P) & s_y(P) \\ t_x(P) & t_y(P) \end{vmatrix} = \begin{vmatrix} 0 & 0 \\ t_x(P) & t_y(P) \end{vmatrix} = 0,$$

which is impossible. □

Recall that

$$w_{st} = 0 \text{ on } \mathbf{R}^2 \iff w = w(s,t) = f(s) + g(t), \tag{8.2.5}$$

where $f, g \in C^2(\mathbf{R})$ are twice continuously differentiable functions (see Example 1.4.3).

Corollary 8.2.2. *Let $a, b, c \in \mathbf{R}$ be constants such that $a \neq 0$ and*

$$b^2 - ac > 0,$$

and let

$$au_{xx} + 2bu_{xy} + cu_{yy} = d(x, y, u, u_x, u_y), \tag{8.2.6}$$

where $d \in C(\mathbf{R}^5)$, be a second-order semilinear (hyperbolic) PDE on $\mathbf{R}^2$. Then:
(i) the standard change of variables, described in Proposition 8.2.1, is

$$\begin{cases} s = y + D_1 x, \\ t = y + D_2 x, \end{cases}$$

where the real numbers D_1, D_2 are defined by

$$D_1 = \frac{-b + \sqrt{b^2 - ac}}{a} \quad \text{and} \quad D_2 = \frac{-b - \sqrt{b^2 - ac}}{a},$$

and this change of variables reduces the PDE (8.2.6) to the PDE

$$w_{st} = d^*(s, t, w, w_s, w_t)$$

in the canonical form, where

$$w(s, t) = u(x, y)$$

and

$$d^*(s, t, w, w_s, w_t) = \frac{a}{4(ac - b^2)} d(x, y, u, u_x, u_y);$$

(ii) *the general solution of the hyperbolic PDE*

$$au_{xx} + 2bu_{xy} + cu_{yy} = 0$$

on $\mathbf{R}^2$ *is therefore*

$$u(x, y) = f(y + D_1 x) + g(y + D_2 x),$$

where $f, g \in C^2(\mathbf{R})$ *are twice continuously differentiable functions.*

Proof. (i) Recalling the proof of Proposition 8.1.1, one easily sees that, by (8.1.3) above,

$$
\begin{aligned}
au_{xx} + 2u_{xy} + cu_{yy} = {} & Aw_{ss} + 2Bu_{st} + Cw_{tt} && \text{(8.2.7)} \\
& + (as_{xx} + 2bs_{xy} + cs_{yy})w_s \\
& + (at_{xx} + 2bt_{xy} + ct_{yy})w_t
\end{aligned}
$$

Clearly,

$$
\begin{aligned}
y' = -D_k \iff {} & y = -D_k x + C \\
\iff {} & y + D_k x = C,
\end{aligned}
$$

where $k = 1, 2$, and so the required first integrals s, t are

$$s(x, y) = y + D_1 x \quad \text{and} \quad t(x, y) = y + D_2 x.$$

Hence the change of variables $x, y \to s, t$ suggested by Proposition 8.2.1, is indeed

$$
\begin{cases}
s = y + D_1 x, \\
t = y + D_2 x,
\end{cases}
$$

as claimed. This change of variables is obviously C^1-invertible on the whole of $\mathbf{R}^2$, since both functions s, t are *linear* and

$$\det \frac{\partial(s, t)}{\partial(x, y)} = \begin{vmatrix} s_x & s_y \\ t_x & t_y \end{vmatrix} = \begin{vmatrix} D_1 & 1 \\ D_2 & 1 \end{vmatrix} = D_1 - D_2 \neq 0.$$

Next, again due to linearity of both s, t, all their second-order partial derivatives are equal to zero:

$$s_{xx} = s_{xy} = s_{yy} = t_{xx} = t_{xy} = t_{yy} = 0.$$

Then, by (8.2.7),

$$au_{xx} + 2bu_{xy} + cu_{yy} = Aw_{ss} + 2Bw_{st} + Cw_{tt}.$$

As we have seen in the proof of Proposition 8.2.1,

$$A = C = 0 \quad \text{and} \quad B = \frac{2}{a}s_y t_y (ac - b^2).$$

Thus

$$au_{xx} + 2bu_{xy} + cu_{yy} = 2Bw_{st} = 4\frac{ac - b^2}{a}w_{st},$$

since

$$s_y = t_y = 1,$$

whence the result follows.

(ii) By (i), under the change of variables $x, y \to s, t$ the equation

$$au_{xx} + 2bu_{xy} + cu_{yy} = 0$$

on $\mathbf{R}^2$ is reduced to the equation

$$w_{st} = 0$$

on $\mathbf{R}^2$.

Finally, we obtain, by (8.2.5), that

$$u(x,y) = /\!/ \quad w(s,t) = f(s) + g(t) \quad /\!/$$
$$= f(y + D_1 x) + g(y + D_2 x),$$

where $f, g \in C^2(\mathbf{R})$, is the general solution of the equation

$$au_{xx} + 2bu_{xy} + cu_{yy} = 0$$

on $\mathbf{R}^2$ as claimed. $\qquad\square$

Example 8.2.1. Reduce the (hyperbolic) equation

$$u_{xx} + u_{xy} - 2u_{yy} + 1 = 0$$

to canonical form, and then find its general solution on $\mathbf{R}^2$.

Solution. The coefficients of the equation in question, as a second-order semilinear equation

$$a(x,y)u_{xx} + 2b(x,y)u_{xy} + c(x,y)u_{yy} = d(x,y,u,u_x,u_y),$$

are

$$a = 1, \quad b = \frac{1}{2}, \quad c = -2, \quad d = -1,$$

and the discriminant of the equation is positive:

$$b^2 - ac = \frac{1}{4} - 1(-2) = \frac{9}{4}.$$

Thus the equation is hyperbolic on $\mathbf{R}^2$. Further,

$$D_1 = \frac{-b + \sqrt{b^2 - ac}}{a} = 1 \quad \text{and}$$
$$D_2 = \frac{-b - \sqrt{b^2 - ac}}{a} = -2.$$

By Corollary 8.2.2, the standard change of variables which reduces the equation in question to canonical form is

$$\begin{cases} s = y + D_1 x, \\ t = y + D_2 x, \end{cases}$$

and it is, in our case,

$$\begin{cases} s = y + x, \\ t = y - 2x. \end{cases} \tag{8.2.8}$$

Now there are two ways to proceed:
(a) we can either write down and then solve the equation

$$w_{st} = d^*,$$

where

$$d^* = \frac{a}{4(ac - b^2)} d,$$

to which, by Corollary 8.2.2, our equation is reduced under the change of variables $x, y \rightarrow s, t$;

(b) or, alternatively, we can first solve the homogeneous equation

$$u_{xx} + u_{xy} - 2u_{yy} = 0$$

whose general solution is also given in Corollary 8.2.2, and then obtain a particular solution of the inhomogeneous equation

$$u_{xx} + u_{xy} - 2u_{yy} = -1$$

and, thereafter, the general solution.

We will explore both ways.

(a) For the problem at hand, we have

$$d^* = \frac{a}{4(ac - b^2)} d = \frac{1}{9},$$

and so the reduced equation is

$$w_{st} = \frac{1}{9}. \tag{8.2.9}$$

The general solution of the corresponding homogeneous equation on $\mathbf{R}^2$ is given by

$$w_{st} = 0 \iff w = w_{\text{hom}}(s, t) = f(s) + g(t),$$

where $f, g \in C^2(\mathbf{R})$ are twice continuously differentiable functions. An evident particular solution of the inhomogeneous equation (8.2.9) is

$$w = w_{\text{prt}}(s, t) = \frac{st}{9}.$$

Thus

$$w_{st} = \frac{1}{9} \iff w = w_{\text{prt}} + w_{\text{hom}} = \frac{st}{9} + f(s) + g(t).$$

The general solution of the original equation on $\mathbf{R}^2$ is therefore

$$u(x, y) = \mathbin{/\!/} \quad w(s, t) = \frac{st}{9} + f(s) + g(t) \quad \mathbin{/\!/} \tag{8.2.10}$$

$$= \frac{(x + y)(y - 2x)}{9} + f(x + y) + g(y - 2x),$$

where $f, g \in C^2(\mathbf{R})$ are twice continuously differentiable functions.

(b) As outlined above, we first solve the homogeneous equation

$$u_{xx} + u_{xy} - 2u_{yy} = 0,$$

using Corollary 8.2.2:

$$u_{xx} + u_{xy} - 2u_{yy} = 0 \qquad \Longleftrightarrow$$
$$u = u_{\text{hom}} = f(y + D_1 x) + g(y + D_2 x) \Longleftrightarrow$$
$$u = u_{\text{hom}} = f(y + x) + f(y - 2x).$$

Observe that the right-hand side of the original equation is a constant one, and so, by a well-known trick, we can look for particular solutions in one variable only. Say, let us look for a particular solution of the form $u = u_{\text{prt}} = z(x)$:

$$u_{xx} + u_{xy} - 2u_{yy} + 1 = 0 \qquad \Rightarrow$$
$$[z(x)]_{xx} + [z(x)]_{xy} - 2[z(x)]_{yy} + 1 = 0 \quad \Rightarrow$$
$$z''(x) + 1 = 0 \qquad \Rightarrow$$
$$z''(x) = -1 \text{ (integrate both parts twice)} \Rightarrow$$
$$z(x) = -\frac{x^2}{2} + C_1 x + C_2.$$

Thus, for instance, the function

$$u_{\text{prt}}(x, y) = -\frac{x^2}{2}$$

is a particular solution of our equation. Hence

$$u_{xx} + u_{xy} - 2u_{yy} = 1 \qquad \Longleftrightarrow$$
$$u = u_{\text{prt}} + u_{\text{hom}} \qquad \Longleftrightarrow$$
$$u = -\frac{x^2}{2} + f(y + x) + g(y - 2x),$$

where f, g are twice continuously differentiable functions on $\mathbf{R}$ (at the first glance, the general solution in the latter form may appear to be different from that one in (8.2.10), but, as it is easy to see, both define the same family of functions). $\square$

The following query to $\mathcal{M}|\alpha$

```
D[ u(x,y), x,x ] +D[ u(x,y), x,y ] -2*D[ u(x,y), y,y ] +1=0
```

produces

$$u(x, y) = c_1(y - 2x) + c_2(x + y) - x^2/2.$$

Sometimes it is desirable to transform/reduce a hyperbolic PDE into its *second* canonical form. To move from the first to the second canonical form, we perform one more change of variables.

Claim 8.2.3. *The change of variables*

$$\begin{cases} s = x + y, \\ t = x - y \end{cases}$$

transforms a hyperbolic PDE

$$u_{xy} = d(x, y, u, u_x, u_y) \tag{8.2.11}$$

to a PDE of the form

$$w_{ss} - w_{tt} = d^*(s, t, w, w_s, w_t),$$

which is called the second canonical form *of the hyperbolic equations.*

Proof. Starting with the basic functional equation

$$u(x, y) = w(s, t),$$

we differentiate it by x, thereby getting that

$$u_x = w_s s_x + w_t t_x = w_s + w_t,$$

since

$$s_x = t_x = [x + y]_x = [x - y]_x = 1.$$

Next,

$$\begin{aligned} u_{xy} = [u_x]_y = [w_s]_y + [w_t]_y \\ = w_{ss} s_y + \cancel{w_{st} t_y} + \cancel{w_{ts} s_y} + w_{tt} t_y \\ = w_{ss} - w_{tt}, \end{aligned}$$

since

$$s_y = [x + y]_y = 1 \quad \text{and} \quad t_y = [x - y]_y = -1.$$

$\square$

8.3 Parabolic equations

Proposition 8.3.1 (Canonical form of the parabolic equations). *Let*

$$au_{xx} + 2bu_{xy} + cu_{yy} = d(x, y, u, u_x, u_y) \tag{8.3.1}$$

be a second-order semilinear equation on a domain Ω in $\mathbf{R}^2$, let the function a be nonzero everywhere on Ω, and let

$$b^2 - ac = 0$$

everywhere on Ω. Let further the function

$$D = \frac{-b}{a}$$

be continuously differentiable on Ω.

(i) *Suppose that for a point P of Ω there exists a subdomain Ω_P of Ω, which contains P, such that*

(a) *the ODE*

$$y' = -D$$

possesses a first integral $s(x,y)$ on Ω_P such that

(b) *the partial derivative s_y is nonzero everywhere on Ω_P.*

Then there is a subdomain Π_P of Ω which contains the point P such that the change of variables

$$\begin{cases} s = s(x,y), \\ t = x \end{cases} \tag{8.3.2}$$

reduces the PDE (8.3.1) on Π_P to a PDE of the form

$$w_{tt} = d^*(s,t,w,w_s,w_t), \tag{8.3.3}$$

called the canonical form *of the parabolic PDEs.*

(ii) *Accordingly, if the ODE*

$$y' = D$$

has a first integral $s(x,y)$ on the whole of Ω such that the change of variables (8.3.2) is C^1-invertible on Ω, then under this change of variables the PDE (8.3.1) on Ω is reduced to a PDE in the canonical form (8.3.3).

Proof. As was discussed above, the function D satisfies the functional equation

$$aD^2 + 2bD + c = 0$$

(the functions D_1 and D_2, constructed formally as in the formulation of Proposition 8.2.1, are now merged into one, since the discriminant is everywhere zero on Ω). Observe for the future use that

$$aD + b = 0 \quad \left(\Longleftrightarrow \quad D = -\frac{b}{a} \right).$$

Exactly as in the proof of Proposition 8.2.1, we obtain, using Proposition 4.1.1, that

$$s_x - Ds_y = 0$$

and then, since $s_y \neq 0$ everywhere on Ω_P,

$$\frac{s_x}{s_y} = D$$

at all points of Ω_P.

Further, we see that the Jacobian determinant of the change of variables $x, y \to s, t$,

$$\det \frac{\partial(s,t)}{\partial(x,y)} = \begin{vmatrix} s_x & s_y \\ t_x & t_y \end{vmatrix} = \begin{vmatrix} s_x & s_y \\ 1 & 0 \end{vmatrix} = -s_y \neq 0$$

is nonzero everywhere on Ω_P, and so the change of variables is C^1-invertible on a suitable subdomain Π_P of Ω, which contains the point P.

Now the coefficient A of w_{ss} in the principal part of the transformed equation is equal to zero:

$$A = as_x^2 + 2bs_x s_y + cs_y^2$$

$$= s_y^2 \left(a\left(\frac{s_x}{s_y}\right)^2 + 2b\left(\frac{s_x}{s_y}\right) + c \right)$$

$$= s_y^2 (aD^2 + 2bD + c)$$

$$= 0.$$

Thus the coefficient A disappears. What is nice is that the coefficient B *also* disappears. Indeed, since for the second function

$$t = t(x,y) = x,$$

participating in the change of variables $x, y \to s, t$, we have that

$$t_x = 1 \quad \text{and} \quad t_y = 0 \tag{8.3.4}$$

we get

$$B = as_x t_x + b(s_x t_y + s_y t_x) + cs_y t_y$$

$$= as_x + bs_y$$

$$= s_y \left(a\frac{s_x}{s_y} + b \right)$$

$$= s_y \underbrace{(aD + b)}_{0}$$

$$= 0.$$

Also, again by (8.3.4),

$$C = at_x^2 + 2bt_xt_y + ct_y^2 = at_x^2 = a \neq 0.$$

Thus the equation

$$Aw_{ss} + 2Bw_{st} + Cw_{tt} = d(x, y, u, u_x, u_y)$$

is in fact the equation

$$aw_{tt} = d(x, y, u, u_x, u_y),$$

which, after dividing both parts by $a = a(x, y)$,

$$w_{tt} = \frac{d(x, y, u, u_x, u_y)}{a(x, y)} = d^*(s, t, w, w_s, w_t)$$

becomes the canonical equation

$$w_{tt} = d^*(s, t, w, w_s, w_t).$$

(ii) In order to apply part (i) of the proposition, we have to show that

$$s_y \neq 0$$

everywhere on Ω. This immediately follows from the assumption that the change of variables $x, y \rightarrow s, t$ is C^1-invertible on Ω, which gives us that

$$\det \frac{\partial(s, t)}{\partial(x, y)} \neq 0,$$

and hence

$$\begin{vmatrix} s_x & s_y \\ 1 & 0 \end{vmatrix} = -s_y \neq 0$$

at all points of Ω. □

As before, let us solve the second-order linear homogeneous parabolic PDE having the canonical form.

Example 8.3.1. For the PDE $w_{tt} = 0$ on $\mathbf{R}^2$, we have that

$$w_{tt} = 0 \iff w = w(s, t) = tf(s) + g(s),$$

where $f, g \in C^2(\mathbf{R})$ are twice continuously differentiable functions.

Solution. Indeed, we integrate both parts of the equation

$$[w_t(s,t)]_t = 0$$

by t, thereby getting that

$$w_t(s,t) = [w(s,t)]_t = f(s),$$

where f is a continuously differentiable function.

Integrating by t once again, that is, integrating

$$[w(s,t)]_t = f(s)$$

by t, we obtain that

$$w(s,t) = tf(s) + g(s),$$

where g is a continuously differentiable function. Taking into account that, by the definition, any solution of a second-order PDE must be *twice* continuously differentiable, we conclude that

$$w(s,t) = tf(s) + g(s),$$

where f, g are twice continuously differentiable on **R**.

Conversely, it is easy to see that any function of the form

$$w(s,t) = tf(s) + g(s)$$

satisfies the PDE

$$w_{tt} = 0. \qquad\qquad \square$$

Corollary 8.3.2. *Let $a, b, c \in$ **R** be constants such that $a \neq 0$ and*

$$b^2 - ac = 0,$$

and let

$$au_{xx} + 2bu_{xy} + cu_{yy} = d(x, y, u, u_x, u_y), \qquad\qquad (8.3.5)$$

where $d \in C(\mathbf{R}^5)$, be a second-order semilinear (parabolic) PDE on $\mathbf{R}^2$.
Then:

(i) The change of variables, described in Proposition 8.3.1, which reduces the equation (8.3.5) to its canonical form, is

$$\begin{cases} s = y - \dfrac{b}{a}x, \\ t = x, \end{cases}$$

and the corresponding reduced equation is

$$w_{tt} = d^*(s, t, w, w_s, w_t),$$

where

$$w(s, t) = u(x, y)$$

and

$$d^*(s, t, w, w_s, w_t) = \frac{1}{a} d(x, y, u, u_x, u_y).$$

(ii) *In effect, the general solution of the second-order parabolic equation*

$$au_{xx} + 2bu_{xy} + cu_{yy} = 0$$

with constant coefficients on $\mathbf{R}^2$ *is given by*

$$u(x, y) = xf(y + Dx) + g(y + Dx)$$

$$= xf\left(y - \frac{b}{a}x\right) + g\left(y - \frac{b}{a}x\right),$$

where the real number D *is given by*

$$D = -\frac{b}{a}$$

and where $f, g \in C^2(\mathbf{R})$ *are twice continuously differentiable functions.*

Proof. (i) We have to recover a first integral of the ODE

$$y' = -D. \tag{8.3.6}$$

Now as D is a real number,

$$y' = \frac{b}{a} \iff y = \frac{b}{a}x + C.$$

Accordingly, we can use the first integral

$$s(x, y) = y - \frac{b}{a}x$$

of the ODE (8.3.6) on $\mathbf{R}^2$. It is then easy to see that the function $s(x, y)$ satisfies the condition of part (ii) of Proposition 8.3.1. Hence the required change of variables $x, y \to s, t,$

$$\begin{cases} s = s(x, y), \\ t = x \end{cases}$$

is as follows:

$$\begin{cases} s = y - \dfrac{b}{a}x, \\ t = x. \end{cases}$$

We see that, as in the proof of the corresponding statement for the hyperbolic equations with constant coefficients, the change of variables is linear, and is C^1-invertible on $\mathbf{R}^2$, since its Jacobian determinant is nonzero on $\mathbf{R}^2$:

$$\det \frac{\partial(s,t)}{\partial(x,y)} = \begin{vmatrix} s_x & s_y \\ t_x & t_y \end{vmatrix} = \begin{vmatrix} -b/a & 1 \\ 1 & 0 \end{vmatrix} = -1 \neq 0.$$

Consequently, by part (ii) of Proposition 8.3.1, the PDE (8.3.5) is reduced to the equation

$$w_{tt} = d^*(s, t, w, w_s, w_t),$$

where d^* is as defined above.

(ii) By recalling (8.2.7) and keeping in mind that the change of variables described in (i) is linear, we obtain that the PDE

$$au_{xx} + 2bu_{xy} + cu_{yy} = 0 \tag{8.3.7}$$

is transformed to the PDE

$$aw_{tt} = 0,$$

which is equivalent to the PDE

$$w_{tt} = 0,$$

since $a \neq 0$.

By Example 8.3.1, the general solution of the latter PDE on $\mathbf{R}^2$ is

$$w_{tt} = 0 \iff w(s, t) = tf(s) + g(s),$$

and then the general solution of the PDE (8.3.7) on $\mathbf{R}^2$ is

$$u(x, y) = /\!/ \quad w(s, t) = tf(s) + g(s) \quad /\!/$$
$$= xf\left(y - \frac{b}{a}x\right) + g\left(y - \frac{b}{a}x\right),$$

where $f, g \in C^2(\mathbf{R})$ are twice continuously differentiable functions. $\qquad\qquad \square$

Example 8.3.2. Reduce the (parabolic) equation

$$4u_{xx} - 12u_{xy} + 9u_{yy} = 5$$

to canonical form and then find its general solution on $\mathbf{R}^2$.

Solution. The coefficients of the equation in question are

$$a = 4, \quad b = -6, \quad c = 9, \quad d = 5.$$

Hence

$$b^2 - ac = 36 - 36 = 0$$

and the equation is parabolic. Next,

$$D = -\frac{b}{a} = \frac{3}{2}.$$

Thus, according to Corollary 8.3.2, the general solution of the homogeneous equation

$$4u_{xx} - 12u_{xy} + 9u_{yy} = 0$$

is

$$u = u_{\text{hom}}(x, y) = xf(y + Dx) + g(y + Dx)$$

$$= xf\left(y + \frac{3}{2}x\right) + g\left(y + \frac{3}{2}x\right).$$

Next, we look for a particular solution $u = u_{\text{prt}}(x, y)$ of our equation of the form $z(y)$ (the form $z(x)$ can be used as well, of course):

$$4u_{xx} - 12u_{xy} + 9u_{yy} = 5 \qquad \Rightarrow$$
$$[z(y)]_{xx} - 12[z(y)]_{xy} + 9[z(y)]_{yy} = 5 \Rightarrow$$
$$z''(y) = \frac{5}{9} \qquad \Rightarrow$$
$$z(y) = \frac{5}{18}y^2 + C_1 y + C_2;$$

Now, for instance, the function

$$u_{\text{prt}}(x, y) = \frac{5}{18}y^2$$

is a particular solution of the PDE under consideration.

Summing up, we see that the general solution of our equation is

$$u = u(x, y) = u_{\text{prt}}(x, y) + u_{\text{hom}}(x, y)$$

$$= \frac{5}{18}y^2 + xf\left(y + \frac{3}{2}x\right) + g\left(y + \frac{3}{2}x\right),$$

where f, g are twice continuously differentiable functions. $\qquad\square$

8.4 Elliptic equations

When dealing with a hyperbolic (resp., parabolic) second-order semilinear PDE

$$au_{xx} + 2bu_{xy} + cu_{yy} = d(x, y, u, u_x, u_y) \tag{8.4.1}$$

on a domain Ω in $\mathbf{R}^2$, where $a \neq 0$ everywhere on Ω, we saw that in order to find a change of variables $x, y \to s, t$, under which the PDE is locally reducible to canonical form, we have to solve a PDE of the form

$$U_x - D_{1,2}U_y = 0$$

on Ω, where

$$D_{1,2} = \frac{-b \pm \sqrt{b^2 - ac}}{a}.$$

Based on experience gained in Chapter 4, we further saw that one can use solutions of the last PDE that are first integrals of the characteristic ODEs

$$y' = -D_{1,2}$$

on Ω.

Now what if the PDE (8.4.1) is elliptic on Ω? Then the above developed theory evidently becomes inapplicable, for the discriminant

$$b^2 - ac$$

of the equation (8.4.1) is strictly negative on Ω, and so the functions $D_{1,2}$ simply do not exist on Ω (as real-valued functions). One of the standard ways out of this deadlock is to use the theory of ODEs for smooth complex-valued functions on $\mathbf{R}$. Recall that a function $f : I \to \mathbf{C}$, where I is an open interval in $\mathbf{R}$ (resp., a function $f : \Omega \to \mathbf{C}$, where Ω is a domain in $\mathbf{R}^2$) is regarded as smooth iff both the real and imaginary parts of f are smooth real-valued functions on I (resp., on Ω). However, the theory ODEs for smooth complex-valued functions on $\mathbf{R}$ is, as one quickly realizes, a rather awkward to apply (for instance, the reader may wish to think how a possible analog of the Peano–Pickard–Lindelöf theorem should be formulated).

On the other hand, things become much more smooth, if we turn to the theory of complex (holomorphic) ODEs. We assume that the reader is familiar with holomorphic functions in one complex variable. Briefly put, a holomorphic function $f(z, w)$ in two complex variables on a domain $\mathcal{O}$ of $\mathbf{C}^2$ is, to use a phrase which may be familiar to the reader from the theory of holomorphic functions in one variable, "locally represented by its Taylor series" everywhere on $\mathcal{O}$.

We refer the reader to Section 8 in Chapter II of [9] to learn some basic properties of holomorphic functions in two variables and the existence and uniqueness theorem for explicit first-order complex ODEs

$$w' = f(z, w)$$

for holomorphic functions $w = w(z)$ in one complex variable, where f is a holomorphic function on a domain $\mathcal{O}$ of $\mathbf{C}^2$ (an analog of the Peano–Pickard–Lindelöf theorem; see [9, pp. 85–86]). This done, the reader should have no difficulty in providing the proof of the following statement (after rereading the proof of Proposition 4.1.1, if necessary).

Proposition 8.4.1. *Let a function $f = f(z, w)$ be holomorphic on a domain $\mathcal{O}$ of $\mathbf{C}^2$. Then the following are equivalent for a holomorphic function $F = F(z, w)$ on $\mathcal{O}$:*
(i) *F is a* first integral *of the complex ODE*

$$w' = f(z, w),$$ (8.4.2)

on $\mathcal{O}$, that is, whenever the graph of a holomorphic solution $w = w(z)$ of the ODE, whose domain is an open disk $\mathcal{D} = \mathcal{D}_w$ of $\mathbf{C}$, is contained in $\mathcal{O}$, then

$$F(z, w(z)) = \text{const}, \quad (z \in \mathcal{D});$$

(ii) *F satisfies the complex PDE*

$$U_z + f(z, w)U_w = 0,$$

on $\mathcal{O}$.

Now we are ready to formulate and prove the main result of the section.

Proposition 8.4.2 (Canonical form of the elliptic equations). *Let*

$$au_{xx} + 2bu_{xy} + cu_{yy} = d(x, y, u, u_x, u_y)$$ (8.4.3)

be a second-order semilinear equation on a domain Ω in $\mathbf{R}^2$, let the function a be nonzero everywhere on Ω, and let

$$b^2 - ac < 0$$

everywhere on Ω. Let further

$$D = D_1 = \frac{-b + \sqrt{b^2 - ac}}{a} = \frac{-b + i\sqrt{ac - b^2}}{a}$$

(a complex-valued function), and let the function $D = D(x, y)$ admit a holomorphic extension $\mathbf{D} = \mathbf{D}(z, w)$ to a domain $\mathcal{O}$ of $\mathbf{C}^2$.

(i) *Suppose that for a point P there exists a subdomain Ω_P of P, which contains P such that*
   (a) *the complex ODE*

$$w' = -\boldsymbol{D}(z, w) \tag{8.4.4}$$

(for holomorphic functions in one complex variable) has a first integral $\boldsymbol{\varphi}(z, w)$ on an open domain $\mathcal{O}_P$ of $\mathbf{C}^2$ satisfying $\Omega_P \subseteq \mathcal{O}_P \subseteq \mathcal{O}$; define then the functions φ, s, t on Ω_P by

$$\varphi(x, y) = \boldsymbol{\varphi}(x, y),$$
$$s(x, y) = \mathrm{Re}(\varphi(x, y)),$$
$$t(x, y) = \mathrm{Im}(\varphi(x, y)),$$

where $\begin{pmatrix} x \\ y \end{pmatrix}$ runs over Ω_P;
   (b) *the partial derivative*

$$\varphi_y = s_y + it_y \neq 0$$

of φ by y is nonzero on the whole of Ω_P.
   Then there is a subdomain Π_P of Ω_P, which contains the point P such that the change of variables

$$\begin{cases} s = s(x, y), \\ t = t(x, y), \end{cases} \tag{8.4.5}$$

reduces the PDE (8.4.3) to a PDE of the form

$$w_{ss} + w_{tt} = d^*(s, t, w, w_s, w_t), \tag{8.4.6}$$

called the canonical form *of the elliptic PDEs.*
   (ii) *Accordingly, if the complex ODE*

$$w' = -\boldsymbol{D}$$

has a holomorphic first integral $\boldsymbol{\varphi}(z, w)$ on a domain $\mathcal{O}$ of $\mathbf{C}^2$ which contains the whole of Ω such that the change of variables (8.4.5), constructed from $\boldsymbol{\varphi}$ as in (i), is C^1-invertible on Ω, then under this change of variables the PDE (8.4.3) on Ω is reduced to a PDE in the canonical form (8.4.6).

Proof. (i) By Proposition 8.4.1,

$$\boldsymbol{\varphi}_z(z, w) - \boldsymbol{D}(z, w)\boldsymbol{\varphi}_w(z, w) = 0$$

everywhere on the domain $\mathcal{O}_P$ of $\mathbf{C}^2$. In particular, taking a point $\begin{pmatrix} x_0 \\ y_0 \end{pmatrix} \in \Omega_P \subseteq \mathcal{O}_P$, we obtain that

$$\boldsymbol{\varphi}_z(x_0, y_0) - D(x_0, y_0)\boldsymbol{\varphi}_w(x_0, y_0) = 0,$$

or

$$\boldsymbol{\varphi}_z(x_0, y_0) - D(x_0, y_0)\boldsymbol{\varphi}_w(x_0, y_0) = 0.$$

Further, the limit

$$\lim_{h \to 0} \frac{\boldsymbol{\varphi}(x_0 + h, y_0) - \boldsymbol{\varphi}(x_0, y_0)}{h}, \tag{8.4.7}$$

where h is a *real* variable, must be, by the definition of partial derivatives of holomorphic functions, equal to $\boldsymbol{\varphi}_z(x_0, y_0)$. On the other hand, the limit in (8.4.7) is equal to the limit

$$\lim_{h \to 0} \frac{\varphi(x_0 + h, y_0) - \varphi(x_0, y_0)}{h}$$
$$= \lim_{h \to 0} \left(\frac{s(x_0 + h, y_0) - s(x_0)}{h} + i \frac{t(x_0 + h, y_0) - t(x_0)}{h} \right),$$

whence we deduce that

$$\boldsymbol{\varphi}_z(x_0, y_0) = \varphi_x(x_0, y_0) = s_x(x_0, y_0) + it_x(x_0, y_0).$$

As, similarly,

$$\boldsymbol{\varphi}_w(x_0, y_0) = \varphi_y(x_0, y_0) = s_y(x_0, y_0) + it_y(x_0, y_0),$$

we conclude that

$$\varphi_x(x_0, y_0) - D(x_0, y_0)\varphi_y(x_0, y_0) = 0$$

and, therefore,

$$\varphi_x - D\varphi_y = 0$$

everywhere on Ω_P.

Next, since, by (b), φ_y is nonzero everywhere on Ω, we obtain that

$$\frac{\varphi_x}{\varphi_y} = D.$$

Now as

$$aD^2 + 2bD + c = 0,$$

we get that

$$a\left(\frac{\varphi_x}{\varphi_y}\right)^2 + 2b\frac{\varphi_x}{\varphi_y} + c = 0,$$

whence

$$a\varphi_x^2 + 2b\varphi_x\varphi_y + c\varphi_y^2 = 0.$$

Assume that the change of variables $x, y \rightarrow s, t$ is C^1-invertible in a suitable sub-domain Π_P of Ω_P which contains the point P (we prove this shortly, by demonstrating, as usual in this chapter, that the Jacobian determinant is nonzero everywhere on Ω_P). Taking then into account that

$$\varphi_x = s_x + it_x \quad \text{and} \quad \varphi_y = s_y + it_y,$$

we obtain that

$$a(s_x + it_x)^2 + 2b(s_x + it_x)(s_y + it_y) + c(s_y + it_y)^2 = 0,$$

or, as it is easy to see,

$$
\begin{aligned}
0 = \quad & as_x^2 + 2ias_xt_x & - at_x^2 + \\
& 2bs_xs_y + 2ibs_xt_y + 2ibt_xs_y & - 2bt_xt_y + \\
& cs_y^2 + 2ics_yt_y & - ct_y^2
\end{aligned}
$$

or

$$(A - C) + 2iB = 0,$$

which implies that

$$A = C \quad \text{and} \quad B = 0$$

everywhere on Π_P. So the change of variables $x, y \rightarrow s, t$ transforms the PDE (8.4.3) on Π_P to the equation

$$Aw_{ss} + Aw_{tt} = d(x, y, u, u_x, u_y),$$

whence, after rewriting x, y, u in terms of s, t, w we obtain the PDE

$$A(w_{ss} + w_{tt}) = d_0(s, t, w, w_s, w_t).$$

Naturally, we wish to divide both parts by A to get the canonical form. So let us show that the new coefficient A $(= C)$ is nonzero everywhere on Π_P.

First, according to our basic statement on change of variables in second-order semilinear equations (Proposition 8.1.1),

$$B^2 - AC = -A^2 = (b^2 - ac)\left(\det \frac{\partial(s,t)}{\partial(x,y)}\right)^2.$$

The discriminant $b^2 - ac$ is clearly nonzero everywhere on Ω (by the conditions). So what is left to show is that the Jacobian determinant is also nonzero everywhere on Ω_P (which, in turn, will imply local C^1-invertibility of our change of variables near P, as we also wish to demonstrate).

We start with

$$\frac{\varphi_x}{\varphi_y} = D \Rightarrow \frac{s_x + it_x}{s_y + it_y} = -\frac{b}{a} + i\frac{\sqrt{ac - b^2}}{a}. \tag{8.4.8}$$

Clearly,

$$\frac{s_x + it_x}{s_y + it_y} = \frac{1}{s_y^2 + t_y^2}(s_x + it_x)(s_y - it_y)$$

$$= \frac{1}{s_y^2 + t_y^2}(s_x s_y + t_x t_y - i(s_x t_y - s_y t_x)).$$

Equating the imaginary parts of both sides of the second equation in (8.4.8) gives us then that

$$-\frac{s_x t_y - s_y t_x}{s_y^2 + t_y^2} = \frac{\sqrt{ac - b^2}}{a}. \tag{8.4.9}$$

Now as the Jacobian determinant is equal to

$$\begin{vmatrix} s_x & t_x \\ s_y & t_y \end{vmatrix} = s_x t_y - s_y t_x$$

on Ω_P, it is, by (8.4.9), nonzero everywhere on Ω_P. This completes the proof of the fact that $A \neq 0$ everywhere on Π_P.

Going back to the transformed equation

$$A(w_{ss} + w_{tt}) = d_0(s, t, w, w_s, w_t),$$

we now can, with a good conscience, divide both sides by A, thereby getting the desired canonical form

$$w_{ss} + w_{tt} = d^*(s, t, w, w_s, w_t),$$

where

$$d^*(s, t, w, w_s, w_t) = \frac{d_0(s, t, w, w_s, w_t)}{A}.$$

(ii) In view of (i), it suffices to show that

$$\varphi_y \neq 0,$$

where φ is the restriction of the first integral $\boldsymbol{\varphi}$ on Ω, everywhere on Ω. It is certainly true, since otherwise

$$s_y(Q) = t_y(Q) = 0$$

at some point Q of Ω, which would imply that the Jacobian determinant

$$\begin{vmatrix} s_x & t_x \\ s_y & t_y \end{vmatrix}$$

of the change of variables $x, y \to s, t$ vanishes at Q. $\qquad\square$

Let us consider an example in which we learn how to construct the standard change of variables under which a given elliptic PDE is locally reducible to canonical form.

Example 8.4.1. Find the standard change of variables described in Proposition 8.4.2 under which the PDE

$$u_{xx} - 2xu_{xy} + (x^2 + y^2)u_{yy} = 0,$$

is locally reducible to canonical form everywhere on the open upper half-plane $\Omega = \mathbf{R} \times \mathbf{R}^+$ in $\mathbf{R}^2$.

Solution. The coefficient functions are

$$a(x, y) = 1, \quad b(x, y) = -x, \quad c(x, y) = x^2 + y^2, \quad d(x, y) = 0,$$

and hence the discriminant of the equation is equal to

$$\Delta(x, y) = b(x, y)^2 - a(x, y)c(x, y) = x^2 - x^2 - y^2 = -y^2.$$

So $\Delta(x, y)$ is negative everywhere on $\Omega = \mathbf{R} \times \mathbf{R}^+$, and the equation is elliptic on Ω.
Next,

$$D(x, y) = \frac{-b(x, y) + i\sqrt{|\Delta(x, y)|}}{a(x, y)} = x + i|y| = x + iy$$

for all $\begin{pmatrix} x \\ y \end{pmatrix} \in \Omega$. The function $D(x, y)$ on Ω evidently admits a holomorphic extension $\boldsymbol{D}(z, w)$ on the whole of $\mathcal{O} = \mathbf{C}^2$, namely, the function

$$\boldsymbol{D}(z, w) = z + iw, \quad \begin{pmatrix} z \\ w \end{pmatrix} \in \mathbf{C}^2.$$

Now consider the complex ODE

$$w' = -\boldsymbol{D}(z, w),$$

that is, the complex ODE

$$w' = -z - iw. \tag{8.4.10}$$

This ODE is a linear inhomogeneous first-order complex ODE, for which the holomorphic function

$$w_{\mathrm{prt}}(z) = iz - 1$$

is a quite evident particular solution. Since, further, the general solution of the corresponding homogeneous equation is given by

$$w' = -iw \iff w = w_{\mathrm{hom}} = Ke^{-iz},$$

we obtain that

$$w' = -z - iw \iff w = w_{\mathrm{prt}} + w_{\mathrm{hom}} = iz - 1 + Ke^{-iz}$$
$$\iff (w - iz + 1)e^{iz} = K,$$

where K is an arbitrary complex constant. The last equivalence demonstrates that the function

$$\boldsymbol{\varphi}(z, w) = (w - iz + 1)e^{iz}, \quad \begin{pmatrix} z \\ w \end{pmatrix} \in \mathbf{C}^2$$

is a first integral of the characteristic ODE (8.4.10) on $\mathcal{O} = \mathbf{C}^2$. An advice to the reader: both $W|\alpha$ and Maple are capable of handling complex ODEs; for instance, we could have used Maple's command

```
dsolve( diff(w(z),z) = -z - I*w(z), w(z));
```

to obtain the general solution of the characteristic ODE (note the symbol I, a Maple *constant*, which is reserved for the imaginary unit i).

The restriction $\varphi(x, y)$ of the function $\boldsymbol{\varphi}(z, w)$ on Ω is therefore the function

$$\varphi(x, y) = (y - ix + 1)e^{ix} = (y - ix + 1)(\cos(x) + i\sin(x)),$$

where $\begin{pmatrix} x \\ y \end{pmatrix} \in \Omega$ and where we have used Euler's identity

$$e^{ix} = \cos(x) + i\sin(x), \quad (x \in \mathbf{R})$$

to justify the second equality.

One easily sees that

$$s(x,y) = \mathrm{Re}(\,\varphi(x,y)\,) = (y+1)\cos(x) + x\sin(x),$$
$$t(x,y) = \mathrm{Im}(\,\varphi(x,y)\,) = (y+1)\sin(x) - x\cos(x),$$

and that

$$\begin{vmatrix} s_x & s_y \\ t_x & t_y \end{vmatrix} = -y \neq 0,$$

for all $\begin{pmatrix} x \\ y \end{pmatrix} \in \Omega$, which means the change of variables $x, y \to s, t$ is locally invertible everywhere on Ω. So the change of variables $x, y \to s, t$ can be used to obtain a local reduction of the PDE in question near every point of Ω to canonical form. We urge the reader to verify that the coefficients A, B, C of the transformed equation are then as follows:

$$A = C = y^2, \quad B = 0,$$

and so everything works as advertised. $\qquad\qquad\square$

Recall that at similar steps in our study of the hyperbolic and parabolic equations, we have discussed homogeneous linear equations having canonical forms: the equation

$$w_{st} = 0 \quad (\Longleftrightarrow\ w = w(s,t) = f(s) + g(t))$$

on $\mathbf{R}^2$ in the hyperbolic case and the equation

$$w_{tt} = 0 \quad (\Longleftrightarrow\ w = w(s,t) = tf(s) + g(s))$$

on $\mathbf{R}^2$ in the parabolic case.

So we have to discuss the corresponding homogenous linear elliptic equation

$$w_{ss} + w_{tt} = 0,$$

or the equation

$$u_{xx} + u_{yy} = 0$$

having canonical form. This equation is called *Laplace's* equation.

However, one quickly discovers that the task of describing the general solution of Laplace's equation in a reasonably simple form is not as straightforward as in the hyperbolic and parabolic case. In anticipation of difficulty in describing solutions of

Laplace's equation, the first step is to give a name for them: a real-valued function u on an open set Ω in $\mathbf{R}^2$ is called *harmonic* if u is a solution of Laplace's equation

$$u_{xx} + u_{yy} = 0$$

on Ω.

Now if $f : \mathcal{O} \to \mathbf{C}$, where $\mathcal{O}$ is an open set in $\mathbf{C}$, is a holomorphic function, then as it should be known, the real-valued functions

$$u(x,y) = \mathrm{Re}(f(x+iy)) \quad \text{and} \quad v(x,y) = \mathrm{Im}(f(x+iy))$$

on $\mathcal{O}$ (as a subset of $\mathbf{R}^2$), must satisfy the Cauchy–Riemann equations

$$u_x = v_y$$
$$u_y = -v_x.$$

Since the derivative of a holomorphic function on an open set in $\mathbf{C}$ is again holomorphic, both functions u, v are in fact infinitely continuously differentiable on $\mathcal{O}$, and we get that

$$u_{xx} + u_{yy} = [u_x]_x + [u_y]_y = [v_y]_x + [-v_x]_y = 0.$$

A similar argument shows that the function v also satisfies Laplace's equation.

Thus *every* holomorphic function $f : \mathcal{O} \to \mathbf{C}$ produces a pair u, v of real-valued harmonic functions on $\mathcal{O}$. Moreover, if u is a real-valued harmonic function on an open set $\mathcal{O}$ in $\mathbf{R}^2$, then u is locally the real part of a suitable holomorphic function near each point of $\mathcal{O}$, and if $\mathcal{O}$ is *simply connected*, u is the real part of a suitable holomorphic function on the whole of $\mathcal{O}$ (see [1, Chapter IV, Section 3], [4, Chapter VIII, Theorem 3.1]).

That said, let us discuss a *conventional* way all solutions of Laplace's equation can be described; in particular, it is used in math software, and due to that, should be remembered. We present the corresponding argument in the (not particular rigorous) form in which it is often found in the literature. Roughly, the idea is to argue as in the *hyperbolic* case.

The coefficients of Laplace's equation

$$u_{xx} + u_{yy} = 0,$$

are

$$a = 1, \quad b = 0, \quad c = 1, \quad d = 0,$$

and then

$$D_{1,2} = \frac{-b \pm \sqrt{b^2 - ac}}{a} = \pm i.$$

The hyperbolic case suggests a change of variables derived from a first integral the complex characteristic ODE

$$y' = i \iff y = ix + K \iff y - ix = K,$$

and a first integral of the complex characteristic ODE

$$y' = -i \iff y = -ix + K \iff y + ix = K,$$

where K is a complex constant. Thus we can try the change of variables

$$\begin{cases} s = y - ix, \\ t = y + ix, \end{cases} \iff \begin{cases} x = i/2 \cdot (s - t), \\ y = 1/2 \cdot (s + t). \end{cases} \tag{8.4.11}$$

It is a linear, and as demonstrated above, a C^1-invertible change of variables, and hence the equation will be transformed to

$$2Bw_{st} = 0 \iff w_{st} = 0$$

since $B \neq 0$.

As the general solution of $w_{st} = 0$ is $f(s) + g(t)$, we conclude that the general solution of Laplace's equation on $\mathbf{R}^2$ can be presented in the form

$$u(x, y) = f(y - ix) + g(y + ix), \tag{8.4.12}$$

where f, g are .... The query

```
D[ u(x,y), x,x ] + D[ u(x,y), y,y ]=0
```

to $W|\alpha$ produces, as it was hinted above, a similar result:

$$u(x, y) = c_1(y + ix) + c_2(y - ix).$$

One of the many questions that arises out of the study of the argument is about the nature of the functions f, g in (8.4.12) (domains? ranges? class of smoothness?). Another question is whether the argument can be properly formalized. Somewhat surprisingly, there is a wide variety of answers to these questions. Some people would state that it is really hard/awkward to describe precisely the nature of functions f, g in (8.4.12), and some people would state that the argument, while "avoiding technicalities," gives us a valuable insight into complex-valued functions (e. g., can be used to introduce the Cauchy–Riemann equations), and so no further formalization is necessary.

Avoiding these extremities, let us sketch the proof, in support of (8.4.12), that every real-valued harmonic function u on $\mathbf{R}^2$ can be written as

$$u(x, y) = \boldsymbol{f}(x + iy) + \boldsymbol{g}(x - iy),$$

where f, g are holomorphic on $\mathbf{C}$, and so both functions in the right-hand side of (8.4.12) can be regarded as holomorphic (why exactly?). Keeping in mind the well-known identification of $\mathbf{R}^2$ with $\mathbf{C}$, which we have already used above, we can think that u is defined on $\mathbf{C}$. By the results from [1, 4] we have quoted above, u is the real part of a suitable holomorphic function f on the whole of $\mathbf{C}$, which we can write as

$$f(z) + \overline{f(z)} = 2u(z),$$

and hence as

$$f(\overline{z}) + \overline{f(\overline{z})} = 2u(\overline{z})$$

for all $z \in \mathbf{C}$. A complex-valued function f on an open set $\mathcal{O}$ in $\mathbf{C}$ is said to be *antiholomorphic* iff the function $\overline{f}$ is holomorphic on $\mathcal{O}$. Clearly, the complex conjugation

$$\sigma(z) = \overline{z}, \quad (z \in \mathbf{C})$$

is antiholomorphic. It is not difficult to show that the composition of an antiholomorphic and a holomorphic (resp., antiholomorphic) is antiholomorphic (resp., holomorphic) [8, Exercise 5, p. 93]. Accordingly, the function

$$g(z) = \overline{f(\overline{z})} = (\sigma \circ f \circ \sigma)(z), \quad (z \in \mathbf{C})$$

is holomorphic on $\mathbf{C}$. This implies that

$$f(\overline{z}) + g(z) = 2u(\overline{z}),$$

and hence

$$f(z) + g(\overline{z}) = 2u(z)$$

for all $z \in \mathbf{C}$, or

$$\frac{1}{2}f(x + iy) + \frac{1}{2}g(x - iy) = u(x, y), \quad \begin{pmatrix} x \\ y \end{pmatrix} \in \mathbf{R}^2,$$

where both f, g are holomorphic on $\mathbf{C}$, as claimed.

Corollary 8.4.3. *Let $a, b, c \in \mathbf{R}$ be constants such that $a \neq 0$ and*

$$b^2 - ac < 0,$$

let

$$au_{xx} + 2bu_{xy} + cu_{yy} = d(x, y, u, u_x, u_y), \tag{8.4.13}$$

where $d \in C(\mathbf{R}^5)$, *be a second-order semilinear (elliptic) PDE. Write D for the* complex number

$$\frac{-b + i\sqrt{ac - b^2}}{a}.$$

Then:

(i) *the standard change of variables described in Proposition 8.4.2 is given by*

$$\begin{cases} s = y + \mathrm{Re}(D)x, \\ t = \mathrm{Im}(D)x, \end{cases}$$

and it reduces the PDE (8.4.13) *to the PDE*

$$w_{ss} + w_{tt} = d^*(s, t, w, w_s, w_t)$$

in the canonical form, where

$$w(s, t) = u(x, y)$$

and

$$d^*(s, t, w, w_s, w_t) = \frac{a}{ac - b^2} d(x, y, u, u_x, u_y);$$

(ii) *the general solution of the elliptic PDE*

$$au_{xx} + 2bu_{xy} + cu_{yy} = 0$$

with constant coefficients on $\mathbf{R}^2$ *is therefore*

$$u(x, y) = h(y + \mathrm{Re}(D)x, \mathrm{Im}(D)x),$$

where $h = h(s, t)$ *is an arbitrary harmonic function on* $\mathbf{R}^2$.

Proof. The function

$$D = D_1(x, y) = \frac{-b + i\sqrt{ac - b^2}}{a}$$

is a complex constant, and evidently possesses a holomorphic extension to $\mathcal{O} = \mathbf{C}^2$. Next, we have for the characteristic complex ODE that

$$w' = -D \iff w = -Dz + K \iff w + Dz = K,$$

where K is a complex constant. Accordingly, a holomorphic first integral of the characteristic ODE can be taken as

$$\varphi(z, w) = w + Dz.$$

The restriction of the function $\boldsymbol{\varphi}(z, w)$ on $\mathbf{R}^2$ is therefore

$$\varphi(x, y) = y + Dx = y + \text{Re}(D)x + i\,\text{Im}(D)x.$$

Now the functions s, t that participate in the standard change of variables from Proposition 8.4.2 are

$$s(x, y) = \text{Re}(\varphi(x, y)) = y + \text{Re}(D)x$$

and

$$t(x, y) = \text{Im}(\varphi(x, y)) = \text{Im}(D)x.$$

The change of variables $x, y \to s, t$ is linear and C^1-invertible on $\mathbf{R}^2$, since

$$\det \frac{\partial(s, t)}{\partial(x, y)} = \begin{vmatrix} s_x & s_y \\ t_x & t_y \end{vmatrix} = \begin{vmatrix} \text{Re}(D) & 1 \\ \text{Im}(D) & 0 \end{vmatrix} = -\text{Im}(D)$$

$$= -\frac{\sqrt{ac - b^2}}{a} \neq 0.$$

Since further,

$$s_x = \text{Re}(D) = -\frac{b}{a} \quad \text{and} \quad s_y = 1,$$

we get that

$$A = as_x^2 + 2bs_x s_y + cs_y^2$$

$$= a\left(-\frac{b}{a}\right)^2 + 2b\left(-\frac{b}{a}\right) + c$$

$$= \frac{ac - b^2}{a}$$

As, by the proof of Proposition 8.4.2, $C = A$ and $B = 0$, and as the change of variables $x, y \to s, t$ is linear, we obtain that

$$au_{xx} + 2bu_{xy} + cu_{yy} = Aw_{ss} + 2Bw_{st} + Cw_{ss}$$

$$= \frac{ac - b^2}{a}(w_{ss} + w_{tt}),$$

and the result follows.

(ii) Since the change of variables in part (i) is linear, the PDE

$$au_{xx} + 2bu_{xy} + cu_{yy} = 0$$

is reduced to Laplace's equation

$$w_{ss} + w_{tt} = 0$$

on $\mathbf{R}^2$. Therefore, the general solution of the PDE in question on $\mathbf{R}^2$ is given by

$$u(x, y) = // \quad h(s, t) \quad // = h(y + \operatorname{Re}(D)x, \operatorname{Im}(D)x),$$

where $h = h(s, t)$ is an arbitrary harmonic function on $\mathbf{R}^2$. $\qquad\qquad$ □

Example 8.4.2. Use Maple to find and perform the standard change of variables which reduces the (elliptic) PDE

$$u_{xx} + u_{xy} + u_{yy} = 0,$$

on $\mathbf{R}^2$ to canonical form and to find its general solution.

Solution. The Maple program which fulfills the requested tasks can be as follows:

```
restart;

# Coefficients

a := 1: b := 1/2: c := 1: d := 0:

D1 := (-b+I*sqrt(a*c-b^2))/a:

# PDE

PDE := a*diff(u(x, y), x, x)
          +2*b*diff(u(x, y), x, y)
          +c*diff(u(x, y), y, y) = 0:



# Standard change of variables: x,y -> s,t

S := (x,y) -> y + Re(D1)*x:
T := (x,y) -> Im(D1)*x:

print( s=S(x,y), t=T(x,y) );

# Expressions of x,y via s,t

varchange:=solve( {s=S(x,y), t=T(x,y)}, {x,y}):

# Transformed PDE

with(PDEtools):
simplify(dchange(varchange, PDE));
```

Upon execution, the program produces the two lines as follows:

$$s = y - \frac{1}{2}x, \quad t = \frac{1}{2}\sqrt{3}x$$

$$\frac{3}{4}\left(\frac{\partial^2}{\partial s^2}u(s,\,t)\right) + \frac{3}{4}\left(\frac{\partial^2}{\partial t^2}u(s,\,t)\right) = 0$$

The first line describes the standard change of variables, the second the transformed equation (note the variable D1 in the program: the use of the letter D as variable's name is forbidden in Maple, since it is reserved for the differentiation operator). As it should, the PDE in question is therefore reduced to Laplace's equation

$$w_{ss} + w_{tt} = 0,$$

and then its general solution on $\mathbf{R}^2$ is given by

$$u(x,y) = /\!/ \quad h(s,t) \quad /\!/ = h\left(y - \frac{1}{2}x, \frac{\sqrt{3}}{2}x\right),$$

where $h = h(s,t)$ is an arbitrary harmonic function on $\mathbf{R}^2$. □

Exercises

1. Using Proposition 8.1.1 and equation (8.2.7), or otherwise, perform the change of variables

$$\begin{cases} s = y/x, \\ t = xy, \end{cases}$$

in the PDE

$$(4x^4 + 1)x^2 u_{xx} + (-8x^4 + 2)xyu_{xy} + (4x^4 + 1)y^2 u_{yy}$$
$$+ (12x^4 - 1)xu_x + (-1 - 4x^4)yu_y = 0$$

on $\Omega = \mathbf{R}^+ \times \mathbf{R}^+$.

2. Determine type, reduce to canonical form, and then find the general solution of each of the following PDEs on $\mathbf{R}^2$:

$$5u_{xx} + 8u_{xy} + 5u_{yy} = 3,$$
$$u_{xx} + 4u_{xy} + 4u_{yy} = -2,$$
$$3u_{xx} + 8u_{xy} - 3u_{yy} = 1.$$

3. Determine type of the PDE

$$4y^3 u_{xx} + 8xy^2 u_{xy} + (1 + 4x^2)yu_{yy} + (4y^2 - 4x^2 - 1)u_y = 0,$$

on $\Omega = \mathbf{R} \times \mathbf{R}^+$, and then find a change of variables under which the PDE is (locally) reducible to canonical form on Ω. If the change of variables is C^1-invertible on Ω, find the corresponding transformed equation; if not, find the coefficients A, B, C of the principal part of the transformed equation to see that it has the required form (use Maple whenever possible).

4. (*Separation of variables for Laplace's equation*). Let $X(x)$ and $Y(y)$ be twice continuously differentiable functions on $\mathbf{R}$. Set $u(x, y) = X(x)Y(y)$, where $x, y \in \mathbf{R}$, and assume that u is a solution of Laplace's equation.

(i) Assume that X is never zero on an open interval I and Y is never zero on an open interval J in $\mathbf{R}$, and then show that

$$f(x) = \frac{X''(x)}{X^2(x)} = -\frac{Y''(y)}{Y^2(y)} = g(y), \qquad (*)$$

where $\begin{pmatrix} x \\ y \end{pmatrix} \in I \times J$.

(ii) As $(*)$ implies that the functions $f(x), g(y)$, being functions in *distinct* independent variables, are constants, show that either

$$X''(x) = k^2 X(x) \text{ on } I \quad \text{and} \quad Y''(y) = -k^2 Y(y) \text{ on } J,$$

or

$$X''(x) = -k^2 X(x) \text{ on } I \quad \text{and} \quad Y''(y) = k^2 Y(y) \text{ on } J,$$

for a suitable $k \in \mathbf{R}$ (squaring k makes things simpler).

(iii) Solve the ODEs,

$$z''(t) = k^2 z(t) \quad \text{and} \quad z''(t) = -k^2 z(t),$$

on $\mathbf{R}$, where $k \in \mathbf{R}$ is a constant, and use (ii) to obtain the general form of $u = u(x, y)$ on $\mathbf{R}^2$.

A Appendix

A.1 Inverse function theorem

Let

$$\Phi\begin{pmatrix} x \\ y \end{pmatrix} = \begin{pmatrix} f(x,y) \\ g(x,y) \end{pmatrix}, \quad \begin{pmatrix} x \\ y \end{pmatrix} \in \Omega$$

be a continuously differentiable map defined on an open set Ω in $\mathbf{R}^2$, which means that $f, g \in C^1(\Omega)$. Recall that the notion of the *Jacobian matrix* of Φ at a point $\begin{pmatrix} x \\ y \end{pmatrix} \in \Omega$,

$$J_\Phi(x,y) = \frac{\partial(f,g)}{\partial(x,y)} = \begin{pmatrix} f_x(x,y) & f_y(x,y) \\ g_x(x,y) & g_y(x,y) \end{pmatrix},$$

takes the baton from the notion of the matrix of a linear map

$$\Psi\begin{pmatrix} x \\ y \end{pmatrix} = \begin{pmatrix} a_{11}x + a_{12}y \\ a_{21}x + a_{22}y \end{pmatrix} = \begin{pmatrix} a_{11} & a_{12} \\ a_{21} & a_{22} \end{pmatrix}\begin{pmatrix} x \\ y \end{pmatrix},$$

where $a_{ij} \in \mathbf{R}$ are constants, for which we have that

$$J_\Psi(x,y) = \begin{pmatrix} a_{11} & a_{12} \\ a_{21} & a_{22} \end{pmatrix}.$$

The *Jacobian determinant* of Φ at $\begin{pmatrix} x \\ y \end{pmatrix} \in \Omega$ is then

$$\det J_\Phi(x,y) = |J_\Phi(x,y)| = \begin{vmatrix} f_x(x,y) & f_y(x,y) \\ g_x(x,y) & g_y(x,y) \end{vmatrix},$$

or, for short,

$$\det J_\Phi = |J_\Phi| = \begin{vmatrix} f_x & f_y \\ g_x & g_y \end{vmatrix}.$$

It is helpful to remember that

$$\det J_\Phi = \begin{vmatrix} f_x & f_y \\ g_x & g_y \end{vmatrix} = \begin{vmatrix} f_x & g_x \\ f_y & g_y \end{vmatrix},$$

since the determinant of the transpose A^t of a given square matrix A is equal to the determinant of A.

https://doi.org/10.1515/9783110677256-009

Suppose that a map

$$\Phi\begin{pmatrix} x \\ y \end{pmatrix} = \begin{pmatrix} f(x,y) \\ g(x,y) \end{pmatrix}$$

on $\mathbf{R}^2$ is differentiable at a point $P \in \mathbf{R}^2$, and a map

$$\Psi\begin{pmatrix} s \\ t \end{pmatrix} = \begin{pmatrix} \alpha(s,t) \\ \beta(s,t) \end{pmatrix}$$

is differentiable the point $Q = \Phi(P)$. Clearly,

$$(\Psi \circ \Phi)(P) = \Psi(\Phi(P)) = \begin{pmatrix} \alpha(f(P), g(P)) \\ \beta(f(P), g(P)) \end{pmatrix}$$

$$= \begin{pmatrix} \alpha(Q) \\ \beta(Q) \end{pmatrix},$$

and then, by the chain rule,

$$J_{\Psi \circ \Phi}(P)$$
$$= \begin{pmatrix} \alpha_s(Q)f_x(P) + \alpha_t(Q)g_x(P) & \alpha_s(Q)f_y(P) + \alpha_t(Q)g_y(P) \\ \beta_s(Q)f_x(P) + \beta_t(Q)g_x(P) & \beta_s(Q)f_y(P) + \beta_t(Q)g_y(P) \end{pmatrix},$$

whence

$$J_{\Psi \circ \Phi}(P) = \begin{pmatrix} \alpha_s(Q) & \alpha_t(Q) \\ \beta_s(Q) & \beta_t(Q) \end{pmatrix} \begin{pmatrix} f_x(P) & f_y(P) \\ g_x(P) & g_y(P) \end{pmatrix},$$

or, in other words,

$$J_{\Psi \circ \Phi}(P) = J_\Psi(\Phi(P)) \cdot J_\Phi(P), \tag{A.1.1}$$

where $\cdot$ denotes the matrix multiplication.

Proposition A.1.1. *Let Ω be an open set in $\mathbf{R}^2$, let*

$$\Phi\begin{pmatrix} x \\ y \end{pmatrix} = \begin{pmatrix} S(x,y) \\ T(x,y) \end{pmatrix},$$

where $S, T \in C^1(\Omega)$, be a continuously differentiable map on Ω, and let $\Lambda = \Phi(\Omega)$.
Suppose that Φ is C^1-invertible, that is, there is a continuously differentiable map $\Psi : \Lambda \to \Omega$,

$$\Psi\begin{pmatrix} s \\ t \end{pmatrix} = \begin{pmatrix} X(s,t) \\ Y(s,t) \end{pmatrix},$$

where $X, Y \in C^1(\Lambda)$, *such that*

$$\Psi \circ \Phi = \mathrm{id}_\Omega \quad and \quad \Phi \circ \Psi = \mathrm{id}_\Lambda.$$

Then:

(i) *whenever $U \subseteq \Omega$ is open, the image $\Phi(U)$ of U under Φ is also open;*

(ii) *the Jacobian matrix $J_\Phi(x, y)$ of Φ is nonsingular and the Jacobian determinant*

$$\det J_\Phi(x, y)$$

of Φ is nonzero everywhere on Ω.

Proof. (i) Given an open subset E of $\mathbf{R}^n$, a function $f : E \to \mathbf{R}^m$ is continuous on E if and only if for every open subset V in $\mathbf{R}^m$, the inverse image $f^{-1}(V)$ of V, the set

$$\{x \in E : f(x) \in V\}$$

is also open (see [3, p. 295]). Since the inverse Φ^{-1} of Φ is continuous, we then obtain that the set

$$(\Phi^{-1})^{-1}(U) = \Phi(U)$$

is open for every open subset U of Ω.

(ii) The Jacobian matrix of the identity map id_Ω on Ω,

$$\mathrm{id}_\Omega \begin{pmatrix} x \\ y \end{pmatrix} = \begin{pmatrix} x \\ y \end{pmatrix}, \quad \begin{pmatrix} x \\ y \end{pmatrix} \in \Omega$$

is equal to the identity matrix I at every point $P \in \Omega$:

$$J_{\mathrm{id}_\Omega}(P) = \begin{pmatrix} 1 & 0 \\ 0 & 1 \end{pmatrix} = I.$$

Now as

$$\mathrm{id}_\Omega = \Phi^{-1} \circ \Phi,$$

we obtain, by (A.1.1),

$$I = J_{\mathrm{id}_\Omega} = J_{\Phi^{-1}}(\Phi(P)) \cdot J_\Phi(P)$$

for all points $P \in \Omega$, and the result follows. □

The Jacobian matrix and the Jacobian determinant of a map

$$\Phi : \Omega \to \mathbf{R}^n, \quad (\Omega \subseteq \mathbf{R}^n) \tag{A.1.2}$$

on an open set Ω in the arithmetic space $\mathbf{R}^n$ taking values in $\mathbf{R}^n$ are defined in a way similar to that one above. Suppose that

$$\Phi \begin{pmatrix} x_1 \\ x_2 \\ \vdots \\ x_n \end{pmatrix} = \begin{pmatrix} f_1(x_1, x_2, \ldots, x_n) \\ f_2(x_1, x_2, \ldots, x_n) \\ \vdots \\ f_n(x_1, x_2, \ldots, x_n) \end{pmatrix},$$

where f_i are real-valued functions defined on Ω.

Let then P be a point of Ω at which all functions f_i are differentiable. Then the Jacobian matrix of Φ at P is the matrix

$$J_\Phi(P) = \frac{\partial(f_1, f_2, \ldots, f_n)}{\partial(x_1, x_2, \ldots, x_n)}(P) = \left\| \frac{\partial f_i}{\partial x_j}(P) \right\| = \|[f_i]_{x_j}\|,$$

where i refers to the row and j for the column number of the matrix in the right-hand side, and the Jacobian determinant of Φ at P is

$$\det J_\Phi(P).$$

The reader should be aware that if Φ is differentiable at a point $P \in \Omega$, then the Jacobian matrix of Φ at P is in fact the matrix of the *derivative* $\Phi'(P)$ of Φ at P, the linear map

$$\Phi'(P) \begin{pmatrix} x_1 \\ \vdots \\ x_n \end{pmatrix} = J_\Phi(P) \begin{pmatrix} x_1 \\ \vdots \\ x_n \end{pmatrix}$$

from the vector space $\mathbf{R}^n$ into itself (see [7, Section 9.10]).

It is not difficult to show that analogs of equation (A.1.1) and Proposition A.1.1 hold for maps of the form (A.1.2) (an exercise the reader is encouraged to do).

Theorem A.1.2 (Inverse Function Theorem (IFT)). *Let* $\Phi : \Omega \to \mathbf{R}^n$, *where*

$$\Phi \begin{pmatrix} x_1 \\ \vdots \\ x_n \end{pmatrix} = \begin{pmatrix} f_1(x_1, \ldots, x_n) \\ \vdots \\ f_n(x_1, \ldots, x_n) \end{pmatrix}$$

be a continuously differentiable map on an open set $\Omega \subseteq \mathbf{R}^n$ *and let* $P = \begin{pmatrix} x_1^* \\ \vdots \\ x_n^* \end{pmatrix}$ *be a point of* Ω. *Then the following are equivalent:*

(i) *the derivative $\Phi'(P) : \mathbf{R}^n \to \mathbf{R}^n$ of Φ at P is an invertible linear map;*
(ii) *the Jacobian matrix*

$$\begin{pmatrix} [f_1]_{x_1}(P) & \cdots & [f_1]_{x_n}(P) \\ \vdots & \vdots & \vdots \\ [f_n]_{x_1}(P) & \cdots & [f_n]_{x_n}(P) \end{pmatrix}$$

of Φ at P is invertible (nonsingular);
(iii) *the Jacobian determinant of Φ at P,*

$$\det J_\Phi(x_1^*, \ldots, x_n^*)$$
$$= \det \frac{\partial(f_1, \ldots, f_n)}{\partial(x_1, \ldots, x_n)}(x_1^*, \ldots, x_n^*)$$
$$= \begin{vmatrix} [f_1]_{x_1}(P) & \cdots & [f_1]_{x_n}(P) \\ \vdots & \vdots & \vdots \\ [f_n]_{x_1}(P) & \cdots & [f_n]_{x_n}(P) \end{vmatrix} \neq 0$$

is nonzero;
(iv) *Φ is locally C^1-invertible near the point P, that is, there is an open neighborhood $O_P \subseteq \Omega$ of P such that the restriction $\Phi|_{O_P}$ of Φ on O_P is invertible and continuously differentiable;*
(v) *the change of variables*

$$\begin{cases} s_1 & = f_1(x_1, \ldots, x_n), \\ \vdots & \vdots \qquad \vdots \\ s_n & = f_n(x_1, \ldots, x_n), \end{cases}$$

is locally C^1-invertible in a suitable open neighborhood O_P of P.

The proof of equivalences

$$\text{(i)} \iff \text{(ii)} \iff \text{(iii)}$$

requires only basic linear algebra, and the equivalence

$$\text{(iv)} \iff \text{(v)}$$

is straightforward. For the proof of the hardest of the implications,

$$\text{(i)} \Rightarrow \text{(iv)},$$

we advise the reader to consult [7, Theorem 9.24]. The implication

$$\text{(iv)} \Rightarrow \text{(iii)}$$

follows from the higher dimensional analog of part (ii) of Proposition A.1.1.

The Inverse Function Theorem is an extremely important tool, and it is used in the proofs of a number of important theorems of multivariable calculus/analysis. In the rest of the Appendix, we shall consider some results which are used in Chapters 5 and 6 on the first-order quasilinear PDEs. In all of the cases, we shall provide the proofs of the results under consideration, and we urge the reader to study the proofs carefully. In all of the proofs below, a special emphasis will be put on the technique of change of variables, due to its evident importance for the theory of PDEs.

We start with a particular case of the implicit function theorem which is used in Chapter 6.

Proposition A.1.3 (Implicit function theorem for real-valued functions in three variables). *Consider a real-valued function $F(x, y, z)$ which is continuously differentiable on an open set in $\mathbf{R}^3$ containing a point $P = \begin{pmatrix} x_0 \\ y_0 \\ z_0 \end{pmatrix}$ such that*

$$F(x_0, y_0, z_0) = 0$$

and

$$F_z(x_0, y_0, z_0) \neq 0.$$

Then there exist an open neighborhood Λ of P, an open neighborhood Ω of the point $\tilde{P} = \begin{pmatrix} x_0 \\ y_0 \end{pmatrix}$, and a continuously differentiable function $u \in C^1(\Omega)$, whose graph is contained in Λ, satisfying

$$F(x, y, u(x, y)) = 0$$

everywhere on Ω (existence). Moreover, if

$$F(x, y, v(x, y)) = 0$$

everywhere on Ω for a function $v \in C^1(\Omega)$, whose graph is contained in Λ, then $u = v$ (uniqueness).

Proof. Consider the change of variables

$$\begin{cases} r = x, \\ s = y, \\ t = F(x, y, z) \end{cases}$$

on an open neighborhood of P. Now, since

$$\det \frac{\partial(r, s, t)}{\partial(x, y, z)} = \begin{vmatrix} 1 & 0 & 0 \\ 0 & 1 & 0 \\ F_x & F_y & F_z \end{vmatrix} = F_z,$$

we obtain, by the IFT, that the change of variables above is locally C^1-invertible about P, which implies that it is C^1-invertible in a suitable open ball Λ centered at P, that is,

$$\begin{cases} r = x, \\ s = y, \\ t = F(x,y,z) \end{cases} \quad \Longleftrightarrow \quad \begin{cases} x = r, \\ y = s, \\ z = Z(r,s,t), \end{cases} \tag{A.1.3}$$

everywhere on Λ and on the corresponding open set Λ' in the rst-space, where the function $Z(r,s,t)$ is continuously differentiable on Λ'.

Next, keeping in mind that the point $P = \begin{pmatrix} x_0 \\ y_0 \\ z_0 \end{pmatrix}$ corresponds to the point P' with the rst-coordinates

$$\begin{pmatrix} r_0 \\ s_0 \\ t_0 \end{pmatrix} = \begin{pmatrix} x_0 \\ y_0 \\ F(x_0,y_0,z_0) \end{pmatrix} = \begin{pmatrix} x_0 \\ y_0 \\ 0 \end{pmatrix} = P'$$

in the rst-space, consider an open ball B' of radius $\delta > 0$ centered at the point P' in the rst-space which is contained in Λ'. Then due to invertibility of the change of variables (A.1.3) on Λ',

$$F(r,s,Z(r,s,t)) = t$$

everywhere on Λ', and, in particular,

$$F(r,s,Z(r,s,0)) = 0$$

for all points $\begin{pmatrix} r \\ s \\ 0 \end{pmatrix}$ in the rst-space that are contained in B'.

By recalling that, by the construction, B' is the open ball of radius δ centered at the point $\begin{pmatrix} x_0 \\ y_0 \\ 0 \end{pmatrix}$ of the rst-plane, we can rephrase the above statement by saying that

$$F(x,y,Z(x,y,0)) = 0$$

for all $\begin{pmatrix} x \\ y \end{pmatrix}$ in the open ball Ω of radius δ centered at $\begin{pmatrix} x_0 \\ y_0 \end{pmatrix}$ in the xy-plane. Set then

$$u(x,y) = Z(x,y,0)$$

for all $\begin{pmatrix} x \\ y \end{pmatrix} \in \Omega$. This settles the existence part.

Now let us prove the uniqueness part. Take a point $\begin{pmatrix} x^* \\ y^* \end{pmatrix} \in \Omega$, and consider the function

$$\alpha(t) = F(x^*, y^*, t)$$

whose domain $\{t\}$ is determined by the condition $\begin{pmatrix} x^* \\ y^* \\ t \end{pmatrix} \in \Lambda$. We claim that the function $\alpha(t)$ is one-to-one. Indeed, suppose that

$$\alpha(t_1) = \alpha(t_2) \tag{A.1.4}$$

for some distinct points $t_1, t_2 \in \mathrm{dom}(\alpha)$. For instance, assume that $t_1 < t_2$. As Λ, being an open ball in $\mathbf{R}^3$, is *convex*, for every $t \in [t_1, t_2]$, the point $\begin{pmatrix} x^* \\ y^* \\ t \end{pmatrix}$ is in Λ, whence we get that α is defined and continuously differentiable on $[t_1, t_2]$. Due to (A.1.4), we obtain, by Rolle's theorem, that there is an inner point $t_0 \in (t_1, t_2)$ at which the derivative of α is equal to zero. On the other hand,

$$0 = \alpha'(t_0) = F_z(x^*, y^*, t_0),$$

which contradicts the condition that F_z is nonzero everywhere on Λ.

Now suppose a function $v \in C^1(\Omega)$, whose graph is contained in Λ, is such that

$$F(x, y, v(x, y)) = 0$$

for all $\begin{pmatrix} x \\ y \end{pmatrix} \in \Omega$. But then

$$0 = \underline{F(x, y, u(x, y)) = F(x, y, v(x, y))}$$

and, as we have seen above, it follows that

$$u(x, y) = v(x, y)$$

for all $\begin{pmatrix} x \\ y \end{pmatrix} \in \Omega$, as required. $\qquad\square$

The general case of the implicit function theorem is discussed in [7, pp. 223–227] (strongly recommended especially if the reader is unsure of how the theorem formulated in the general case).

A.2 Functional dependence

Next, we are going to study a number of important corollaries of the Inverse Function Theorem involving continuously differentiable maps

$$\Phi\begin{pmatrix} x \\ y \end{pmatrix} = \begin{pmatrix} f(x, y) \\ g(x, y) \end{pmatrix}$$

whose Jacobian determinant is, quite contrary to the conditions of the IFT, *identically zero* on some nonempty open set in $\mathbf{R}^2$. Note that the material in this section, which is used in Chapter 6, is not always included in the standard syllabi of the multidimensional calculus/analysis, and so there is a high chance that the reader may be unfamiliar with (any of) it.

The Inverse Function Theorem can be viewed (in a way) as a rather dramatic generalization of the following fact from linear algebra: a *linear* map

$$\Phi\begin{pmatrix} x \\ y \end{pmatrix} = \begin{pmatrix} f(x,y) \\ g(x,y) \end{pmatrix} = \begin{pmatrix} a_{11}x + a_{12}y \\ a_{21}x + a_{22}y \end{pmatrix}$$

is invertible if and only if its (Jacobian) matrix

$$A = \begin{pmatrix} a_{11} & a_{12} \\ a_{21} & a_{22} \end{pmatrix}$$

is nonsingular, or, equivalently has a nonzero determinant. What then happens if the determinant of the matrix of Φ *is* equal to zero? Well, if at least one of the maps f, g is not zero, which we can express by saying that rank of the matrix A is equal to one, then the maps are proportional, and hence either

$$f(x,y) = \alpha g(x,y),$$

or

$$g(x,y) = \beta f(x,y)$$

identically on $\mathbf{R}^2$ for some constants $\alpha, \beta \in \mathbf{R}$, and so one of the functions is a function of another: say,

$$f(x,y) = F(g(x,y))$$

where $F(s) = \alpha s$.

Proposition A.2.1. *Suppose that Ω is an open set in $\mathbf{R}^2$ and the Jacobian matrix of a continuously differentiable map*

$$\Phi\begin{pmatrix} x \\ y \end{pmatrix} = \begin{pmatrix} f(x,y) \\ g(x,y) \end{pmatrix},$$

where $f, g \in C^1(\Omega)$, on Ω is of constant rank one on Ω,

$$\operatorname{rank} \frac{\partial(f,g)}{\partial(x,y)} \equiv 1$$

(which means, in effect, that the Jacobian determinant of Φ is identically zero on Ω). Then:

(i) *for every point $P \in \Omega$, there is an open neighborhood O_P of P and a continuously differentiable function $F = F_P(s)$ such that either*

$$f(x,y) \equiv F(g(x,y)), \tag{A.2.1a}$$

or

$$g(x,y) \equiv F(f(x,y)) \tag{A.2.1b}$$

identically on O_P;

(ii) *moreover, (A.2.1a) takes place provided that the gradient vector of g at P,*

$$\nabla g(P) = \begin{pmatrix} g_x(P) \\ g_y(P) \end{pmatrix} \neq \begin{pmatrix} 0 \\ 0 \end{pmatrix},$$

is nonzero and (A.2.1b) takes place provided that the gradient vector $\nabla f(P)$ of f at P is nonzero.

We therefore see that as in the situation with the Inverse Function Theorem, a fact from linear algebra has been generalized to a powerful, but still "local" statement. To provide some more practice in the technique, we shall exploit a change of variables in the proof that follows.

Proof. Let P be a point of Ω. Since the rank of the Jacobian matrix of Φ at P is one, then of one the partial derivatives of f, g is nonzero at P. For instance, assume that

$$g_x(P) \neq 0.$$

Consider then the change of variables

$$\begin{cases} s = g(x,y), \\ t = y. \end{cases} \tag{A.2.2}$$

We have that

$$\det \frac{\partial(s,t)}{\partial(x,y)}(P) = \begin{vmatrix} g_x(P) & g_y(P) \\ 0 & 1 \end{vmatrix} = g_x(P) \neq 0.$$

By the IFT, the change of variables (A.2.2) is then C^1-invertible in some open neighborhood U_P of P which is contained in Ω. In other words, the map

$$\Psi \begin{pmatrix} x \\ y \end{pmatrix} = \begin{pmatrix} g(x,y) \\ y \end{pmatrix}, \quad \begin{pmatrix} x \\ y \end{pmatrix} \in U_P$$

on U_P is C^1-invertible. Write Q for $\Psi(P)$ and U_Q for the open, by Proposition A.1.1, set $\Psi(U_P)$.

Due to C^1-invertibility of our change of variables, there is a continuously differentiable function $F(s, t)$ on U_Q such that

$$f(x, y) = //\ \ F(s, t)\ \ // = F(g(x, y), y)$$

everywhere on U_P. By the chain rule,

$$f_x(x, y) = F_s(g(x, y), y)g_x(x, y) + F_t(g(x, y), y)y_x$$
$$= F_s(g(x, y), y)g_x(x, y),$$

or, for short,

$$f_x = F_s g_x$$

everywhere on U_P. Similarly,

$$f_y = F_s g_y + F_t$$

everywhere on U_P.

Now

$$0 = \det \frac{\partial(f, g)}{\partial(x, y)} = \begin{vmatrix} f_x & f_y \\ g_x & g_y \end{vmatrix} = \begin{vmatrix} F_s g_x & F_s g_y + F_t \\ g_x & g_y \end{vmatrix},$$

whence, after subtracting from the first row of the last determinant the second row multiplied by F_s,

$$0 = \begin{vmatrix} 0 & F_t \\ g_x & g_y \end{vmatrix} = -F_t g_x.$$

Thus

$$F_t g_x = 0$$

everywhere on U_P. Since g_x is nonzero everywhere on U_P, we obtain that

$$F_t(g(x, y), y) = 0$$

everywhere on U_P, or, in other words,

$$F_t(s, t) = 0$$

everywhere on U_Q.

Let then D_Q be an open disk centered at Q which is contained in U_Q. As D_Q is open and convex, and F_t vanishes identically on D_Q, we conclude, by Proposition 1.4.1, that $F(s, t)$ does not depend on the variable t on D_Q. This implies that

$$f(x, y) = F(g(x, y), y) = F(g(x, y))$$

everywhere on the open neighborhood $O_P = \Psi^{-1}(D_Q)$ of P, as claimed.

(ii) The remaining cases

$$g_y(P) \neq 0, \quad f_x(P) \neq 0, \quad f_y(P) \neq 0$$

follow by symmetry. □

Apart from Proposition A.2.1, its three-dimensional analog, which is as follows, is also needed in Chapter 6.

Proposition A.2.2. *Suppose that the Jacobian matrix of a continuously differentiable map*

$$\Phi\begin{pmatrix} x \\ y \\ z \end{pmatrix} = \begin{pmatrix} f(x,y,z) \\ g(x,y,z) \\ h(x,y,z) \end{pmatrix}$$

defined on an open set $\Lambda \subseteq \mathbf{R}^3$ is of constant rank two on Λ,

$$\operatorname{rank} \frac{\partial(f,g,h)}{\partial(x,y,z)} \equiv 2.$$

(Accordingly, the Jacobian determinant of Φ is identically zero on Λ.)

Then:

(i) For every point $P \in \Lambda$, there is an open neighborhood O_P of P and a continuously differentiable function $F = F_P(s,t)$ such that either

$$f(x,y,z) \equiv F(g(x,y,z), h(x,y,z)),$$

or

$$g(x,y,z) \equiv F(f(x,y,z), h(x,y,z)),$$

or

$$h(x,y,z) \equiv F(f(x,y,z), g(x,y,z))$$

identically on O_P.

(ii) In the case when the gradients of f and g,

$$(\nabla f)(P) = \begin{pmatrix} f_x(P) \\ f_y(P) \\ f_z(P)) \end{pmatrix}, \quad (\nabla g)(P) = \begin{pmatrix} g_x(P) \\ g_y(P) \\ g_z(P)) \end{pmatrix},$$

are linearly independent at all points $P \in \Lambda$, we have that

$$h(x,y,z) \equiv F_P(f(x,y,z), g(x,y,z))$$

identically on a suitable neighborhood $O_P \subseteq \Lambda$ of P for a suitable continuously differentiable function $F_P(s,t)$.

If Λ is an open set in $\mathbf{R}^3$, and $f, g \in C^1(\Lambda)$ are such that the gradients of f and g,

$$\nabla f \quad \text{and} \quad \nabla g,$$

are linearly independent everywhere on Λ, or, equivalently if

$$\operatorname{rank} \begin{pmatrix} f_x & f_y & f_z \\ g_y & g_y & g_z \end{pmatrix} \equiv 2$$

everywhere on Λ, then f, g are said to be *functionally independent* on Λ.

Proof of Proposition A.2.2. The proof will go along the same lines as the proof of Proposition A.2.1. Take a point $P \in \Lambda$ and assume, losing no generality in doing so, that

$$\begin{vmatrix} f_x(P) & f_y(P) \\ g_x(P) & g_y(P) \end{vmatrix} \neq 0$$

at P (the other cases, as well as the proof of (ii), will then follow by symmetry). We then consider the change of variables

$$\begin{cases} r = f(x, y, z), \\ s = g(x, y, z), \\ t = z, \end{cases}$$

for which we have that

$$\det \frac{\partial(r, s, t)}{\partial(x, y, z)}(P) = \begin{vmatrix} f_x(P) & f_y(P) & f_z(P) \\ g_x(P) & g_y(P) & g_z(P) \\ 0 & 0 & 1 \end{vmatrix}$$

$$= \begin{vmatrix} f_x(P) & f_y(P) \\ g_x(P) & g_y(P) \end{vmatrix} \neq 0$$

Accordingly, there is an open neighborhood U_P of P on which the corresponding restriction Ψ of the map Φ is C^1-invertible, and the set $U_Q = \Psi(U_P)$ is an open neighborhood of the point $Q = \Psi(P)$.

As Ψ is C^1-invertible, there is a C^1-function $F(r, s, t)$ on U_Q such that

$$h(x, y, z) = F(f(x, y, z), g(x, y, z), z) \tag{A.2.3}$$

everywhere on U_P. It follows that

$$0 = \begin{vmatrix} f_x & f_y & f_z \\ g_x & g_y & g_z \\ h_x & h_y & h_z \end{vmatrix} = \begin{vmatrix} f_x & f_y & f_z \\ g_x & g_y & g_z \\ F_r f_x + F_s g_x & F_r f_y + F_s g_y & F_r f_z + F_s g_z + F_t \end{vmatrix},$$

which implies that

$$\begin{vmatrix} f_x & f_y & f_z \\ g_x & g_y & g_z \\ 0 & 0 & F_t \end{vmatrix} = F_t \begin{vmatrix} f_x & f_y \\ g_x & g_y \end{vmatrix} = 0$$

everywhere on U_P. Consequently,

$$F_t(f(x,y,z), g(x,y,z), z) = 0$$

everywhere on U_P, or, in other words,

$$F_t(r,s,t) = 0$$

everywhere on U_Q. The proof of the following result, an analog of Proposition 1.4.1, is left to the reader.

Lemma. *Suppose that* Π *is a z-simple domain in* $\mathbf{R}^3$ *and*

$$u_z = 0$$

everywhere on Π *for a* C^1*-function* $u = u(x,y,z)$ *on* Π*. Then there exists an* C^1*-function* v *in two variables such that*

$$u(x,y,z) = v(x,y)$$

everywhere on Π*.*

Take an open ball B_Q centered at Q which is contained in U_Q. As, by the lemma, the function F is in fact a function in two variables on B_Q, we obtain from (A.2.3) that

$$h(x,y,z) = F(f(x,y,z), g(x,y,z))$$

at all points of the open neighborhood $\Psi^{-1}(B_Q)$ of the point P. $\qquad\square$

Bibliography

[1] H. Cartan, *Elementary theory of analytic functions of one or several complex variables*, Dover, 1995.

[2] B. R. Gelbaum, J. M. H. Olmsted, *Counterexamples in analysis*, Dover, 2003.

[3] E. Hairer, G. Wanner, *Analysis by its history*. Springer, 2nd edition, 2008.

[4] S. Lang, *Complex analysis*, Springer, 4th edition, 1999.

[5] L. S. Pontryagin, *Ordinary differential equations*, Pergamon Press, 1962.

[6] L. S. Pontryagin, *Ordinary differential equations* (Russian), Nauka, 4th edition, 1974.

[7] W. Rudin, *Principles of mathematical analysis*, McGraw-Hill, 3rd edition, 1976.

[8] M. E. Taylor, *Introduction to complex analysis*, AMS, 2019.

[9] W. Walter, *Ordinary differential equations*, Springer, 1998.

https://doi.org/10.1515/9783110677256-010

Index

Burgers' equation 2, 93, 125

Cauchy problem for a first-order PDE 129
– base curve 130
– equivalence of local solutions 183
– equivalence of solutions 139
– having a unique solution up to equivalence 140
– having infinitely many pairwise nonequivalent solutions 140
– having no solution 130
– initial curve 130
– local solution 140
– solution 129
– solution set 130
change of variables
– C^1-invertible 33
– locally C^1-invertible 39
change of variables in a PDE
– "direct" approach 34
– "indirect" approach 41
Charpit, P. 90

domain
– horizontally convex 24
– vertically convex 24
– x-simple 24
– y-simple 24

first integral
– of an explicit first-order ODE 64
– of the characteristic system of a first-order quasilinear PDE 101
– of the characteristic system of a vector field 101
first-order quasilinear PDE 85
– characteristic curves 92
– characteristic surface 161
– characteristic system 92
– Monge surface 161
first-order semilinear PDE 63
– characteristic curves 68
– characteristic ODE 67, 74
functionally independent functions 105, 261

harmonic function 240

Inverse Function Theorem (IFT) 39, 252

Jacobi, C. G. J. 90
Jacobian determinant 249, 252
Jacobian matrix 249, 252

Lagrange, J.-L. 90
Lagrange–Charpit equations 91
Laplace's equation 2, 239
– separation of variables 247
Lindelöf, E. L. 64

method of characteristics
– for first-order homogeneous linear PDEs 67
– local version 72
– for first-order quasilinear PDEs
– local version 184
– for first-order semilinear PDEs 74
Monge, G. 90, 157, 159

ordinary differential equation (ODE) 1
– complete integral 65
– exact 10
– linear 4
– homogeneous 5
– inhomogeneous 5
– separable 12

partial differential equation (PDE) 1
– first-order quasilinear 85
– first-order semilinear 63
– linear 5
– homogeneous 7
– inhomogeneous 7
– order 2
– reducible to an ODE 10
– second-order semilinear 205
– solution 2
– transformed under a change of variables 36
Peano, G. 64
Peano–Pickard–Lindelöf theorem
– for first-order explicit ODEs 63
– for systems of first-order explicit ODEs 86
Picard, C. E. 64

second-order semilinear PDE 205
– discriminant 205
– elliptic 209
– canonical form 233
– hyperbolic 209

– first canonical form 215
– second canonical form 223
– parabolic 209
– canonical form 224
system of explicit first-order ODEs
– autonomous 90
– nonextendable solution 88

theorem
– Inverse Function (IFT) 39, 252
– on continuous dependence on initial values
 89
– on differentiability with respect to initial
 values 89

– on local solutions of a Cauchy problem for
 first-order quasilinear PDEs 142
– on solution sets of first-order quasilinear PDEs
 124
– on the method of characteristics for first-order
 quasilinear PDEs 175
– Peano–Pickard–Lindelöf 63, 86
transversality condition 142, 182, 184

vector field 91
– characteristic curves 91
– characteristic system 91

www.ingramcontent.com/pod-product-compliance
Lightning Source LLC
Chambersburg PA
CBHW061351210326
41598CB00035B/5953

* 9 7 8 3 1 1 0 6 7 7 2 4 9 *